THE WORKS TRIUMPHS

50 years in motorsport

INCLUDING THE WORKS STANDARDS

Graham Robson

Foulis

Haynes

®

First published in 1993

British Library Cataloguing-in-Publication Data:
A catalogue record for this book is available from the
British Library.

ISBN 0-85429-926-2

Library of Congress catalog card no. 92-75812

G.T. Foulis & Company is a member of the Haynes
Publishing Group P.L.C., Sparkford, Nr. Yeovil,
Somerset, BA22 7JJ.

Printed in Great Britain by The Bath Press, Avon.

THE
WORKS
TRIUMPHS

By the same author

CLASSIC CARS
Collecting, Restoring & Driving

ROVER STORY
(4th edition)

COSWORTH
The search for power
(2nd edition)

THE WORKS ESCORTS
(2nd edition)

LAMBORGHINI COUNTACH

MINI COOPER AND COOPER S SUPER PROFILE

RILEY SPORTS CARS 1926-38

Contents

Acknowledgements

Although I have had Triumph connections for 30 years and more, I still needed a lot of help to make this a truly comprehensive story. There's no doubt that my principal thanks must go to Harry Webster, not only for asking me to run the motorsport operation for him in the 1960s, but for giving me so much of his time, and memory, on other occasions when I was digging deeply into Standard and Triumph history.

As with Harry Webster, so with John Lloyd, who not only managed me so benignly, but gave so much to Standard-Triumph's motorsport effort in the 1950s, 1960s and 1970s.

Now, my thanks to the Triumph team managers over the years who, from time to time, have told me—or refused to tell me—about their time at the helm. The complete list of managers is illustrious—Donald Healey, Ken Richardson, Ray Henderson, 'Kas' Kastner, Peter Browning, Basil Wales, Richard Seth-Smith, Bill Price and John Davenport.

Naturally, I have used interview material which I gathered for previous Triumph books, but for this book I am also grateful to the following for their reminiscences: Keith Ballisat; Gordon Birtwistle; Peter Browning; Brian Culcheth; John Davenport; Roy Fidler; Maurice Gatsonides; Brian Harrocks; 'Kas' Kastner; Jimmy Ray and Johnny Wallwork.

Pictures came from many sources, notably from the old Standard-Triumph archive, and from team drivers themselves, so it is now time for a big Thank You to: T. Kerr Baillie; Keith Ballisat; Gordon Birtwistle; John Brigden; Peter Browning; Anders Clausager (British Motor Industry Heritage Trust); Mike Cook; Cyril Corbishley; Brian Culcheth; John Davenport; Mirco Decet; Roy Fidler; Fred Gallagher; Graham Gauld; Maurice Gatsonides; Geoffrey Healey; 'Kas' Kastner; Richard Langworth; Jimmy Ray; John Sprinzel; Alan Taylor; Jean-Jacques Thuner; Johnny Wallwork; David Weguelin and Alan Zafer.

The following also helped me, with pictures, information or comments: Michael Allen; Peter Garnier; John Hopwood; Martyn King; Tony Merrick; Roger Morris (Standard Motor Club); Gunnar Palm; Bill Piggott; Bill Price; Quadrant Picture Library; Major Tony Rolt and Anders Tunberg.

I am sure all their efforts have made this book bigger, more accurate and more colourful than would otherwise have been possible.

Introduction

This is the complete story of Triumph cars in motorsport, of the factory-supported cars which raced and rallied all round the world. Because the story began with Donald Healey in 1928, and ended with the TR7 V8s in 1980, it covers more than 50 years, and includes many famous victories.

Ever since I started writing about the history of rallying, and rally cars, I have been fascinated by Triumph's involvement in motorsport. Not only were the TR3As, the 2.5 PIs, the Dolomites and the TR7 V8s so glamorous, and so very exciting to drive and watch, but I also had a personal connection. In the 1950s and 1960s I co-drove many a Triumph in many a rally and, in the 1960s I also ran the 'works' Competitions Department in Coventry.

In addition, I wanted to make sure that Standard cars—the other 'arm' of Standard-Triumph—got their due, for in the 1950s the two marques fought side-by-side over the tracks, mountains and special tests of Europe. In looking for honour, I made sure that a Standard Ten was not ignored in favour of a TR3A, or a Pennant in favour of a Herald. The variety and the spread of success was fascinating.

Many years before I started to tackle this job, however, I already knew that it wasn't going to be easy. Over the years the 'works' Triumphs were run from so many different departments, by so many different managers, and controlled by so many directors, that mountains of records were lost, destroyed or dispersed.

By scouring all the magazines, all the books, and all the picture libraries that I could find, it was possible to complete most of the jigsaw, but I didn't want to end up with a 'cuttings job', or a series of second-hand stories.

To be sure that this book tells everything, or as much as records and memories will allow, I decided to interview the personalities who helped create those victories—the drivers, the team managers and the engineers who had been close to the cars.

In many cases they confirmed the stories I had already seen reported, at second hand, but in many cases there were real revelations about the cars, the personalities and the events of the day. I wouldn't have missed a minute of it all. This book, therefore, is by no means a turgid statement of facts, but a series of personal stories, by and about the people who made the 'works' Standards and Triumphs famous.

Chapter 1

Triumph in the 1930s
Inspired by Donald Healey

The first great sporting achievement by a Triumph was neither a success, nor was it achieved by a 'works'-backed car. It was not even an extensively pre-planned factory effort. Although the car was owned by the factory, it was in fact on loan to Donald Healey, and in it he made his epic drive to Monte Carlo in the 1929 Rally, when he failed to reach the final control by a mere two minutes. It was this which started it all.

If Donald Healey, that redoubtable Cornishman, had not chosen Triumph Super Sevens for some of his earliest sporting achievements, and had not picked a Triumph to battle his way through the snows to Monte Carlo, or to use to win Britain's Brighton and Bournemouth rallies, it might have languished for a good many years. The fact is that Triumph's reputation as a sports car, like that of Invicta, was boosted by Donald Healey's skills.

Although Triumph had been a famous builder of pedal cycles since 1890, and of motor cycles since 1902, the first Triumph car was not launched until 1923. At that time the company was still controlled by Siegfried Bettman, a German-born businessman of Jewish descent who had no interest in motor sport in any form. At that time, too, there was no

Below The genius at rest—Donald Healey. A typical study of the man who almost single-handedly transformed Triumph's image in the 1930s. (Geoffrey Healey)

Below right Four of the most important characters in Triumph's mid-1930s motorsport activity (left to right): Jack Ridley (Competitions Manager), Victor Leverett (Sales Manager), Col. Claude Holbrook (Managing Director), and Donald Healey (Technical Director). The car is a Gloria Monte Carlo. (Geoffrey Healey)

connection between Triumph and Standard, except that both were independent concerns which built their cars in Coventry.

The first Triumph cars were built at Clay Lane in Coventry, not far from the city's football ground. The original models were sturdy but simple designs known as the 10/20 types, which used side-valve four-cylinder engines, and which were sold as saloons, tourers, and even as boat-tailed sports cars.

Even though Sir Harry Ricardo had designed the engine, there was little sporty potential in the design, for it had a mere 23.5 bhp to propel a car which, in normal tourer form, weighed around 2,000 lb. Although it was in competition with another new sports car—the 11.9 hp or 14/28 Super Sports MG—there was neither the will nor the equivalent of Cecil Kimber in Coventry to give it a sporting reputation.

However, later in the decade, in 1927, Triumph launched the Super Seven, a much smaller car than the 10/40, which was designed to take on the Austin Seven. A year later it was also faced with another formidable competitor—the original Morris Minor.

Although this tiny Triumph had to work very hard to deliver acceptable performance, it was a willing little machine which was equally as lively and as economical as its rivals. Although it was afflicted with a 832 cc side-valve engine which produced a mere 15 bhp, and was backed by a three-speed gearbox, it had a top speed of 48 mph and could record more than 40 mpg. The Austin Seven and Morris Minor models had a similar performance.

Even though it had virtually no sporting traditions, in 1928 Triumph raised the hopes of several enthusiasts by launching a sports two-seater version with a supercharged engine. This featured a Cozette blower which drew its fuel-air mixture through a Zenith carburettor, could cruise at 55–60 mph, and had a top speed which approached 70 mph. This was a lot more encouraging, for it proved that the engine was tuneable, and reliable when supercharged.

By this time, too, an ambitious young motor trader from Perranporth in Cornwall, Donald Healey, had taken up an agency to sell Triumph cars, and was already entering his own demonstrator models in rallies and trials. It was Healey's first link with Triumph, but this was to become much closer in the 1930s when he joined the firm, eventually to become technical director.

1928-1930

Donald Healey was born in Perranporth in 1898, and seems to have been fascinated by machines, particularly motor cars, from a very early age. Having studied at Newquay College, he was apprenticed to the Sopwith Aviation Company. He soon joined the Royal Flying Corps, learned to fly and won his pilot's wings, then became a bomber pilot over the Western Front in 1916. Soon after this he was shot down, and was sent back to the UK.

Having been invalided out of the service, he eventually returned to his native Cornwall and set up a garage business—the first in Perranporth—trading as 'D.M. Healey— Automobile Engineer'. It wasn't long before he added a chauffeur-driven hire car and coach operation to the thriving little enterprise, and at the age of 24, still as a hobby, he turned his attention to motorsport.

The first car he bought specially for this purpose was an air-cooled ABC which he immediately began to use in the famous British long-distance trials, such as the Exeter and the Lands End. For the Lands End, no less, he 'discovered' the hill-climb at Bluehills Mine, near his home, made sure that the organizers included it in their 1924 event, and used the ABC to win his first Gold Medal for an unpenalized run.

By this time Donald Healey was becoming well-known in the motor trade, and in the motor industry, for he was not only a considerable 'character' but had an attractive personality, a persuasive manner, and a thrusting ambition to get ahead in his businesses. By the mid-1920s he had taken on several agencies, including those of Ariel (from Birmingham) and Riley (from Coventry), and he had already forged a friendship with Gordon Parnell, of the Lockheed brake company. .

The Parnell connection was important to him, for Gordon eventually moved to Triumph in Coventry to supervise the design of cars like the original Super Seven model. This was where the original Healey-Triumph connection was forged, as Donald Healey confirms in his autobiography My World of Cars (Patrick Stephens Ltd, 1989).

> I drove the most unlikely-looking Triumph Seven and Super Seven family saloons in all sorts of events in this country, from local rallies, trials and speed events, to speed trials at Brooklands and the MCC's long distance reliability trials. Though they could scarcely be regarded as 'competition cars', they did well in the Lands End, Exeter and Edinburgh trials, events which attracted entries of 400-plus motorcycles and cars, and included a fair proportion of official, works-entered, cars.'

In fact the first recorded major success for the Healey-Triumph combination was in the Lands End Trial of 1928, where Healey gained another Gold Medal, helped no doubt by his special knowledge of the test hills of his native Cornwall. On that occasion, The Autocar's reporter saw Healey driving a four-seater Triumph Super Seven Tourer up Porlock, and noted that:

> D.M. Healey (Triumph) found no difficulty in making a good start, and his passenger unconcernedly manipulated a baby cinematograph camera as the car passed through the lanes of spectators.

At Bluehills Mine he saw that:

> The first arrival, D.M. Healey (Triumph) charged the bend hell for leather, twirled the steering wheel hard over, and, with a little wheelspin, was round the hairpin and off up the straightforward 1-in-6 section which leads to the summit.

The same issue, of 13 April 1928, showed a photograph of the locally registered car on its way up Porlock hill.

[Not only that, but there was a note that: 'Perranporth and Healey's garage gave everyone—everyone, mark you, and there were more than 500 machines—an excellent tea to soothe frazzled nerves before Bluehills Mine hill. Now that is a good exercise in public relations . . . .']

It was only a small step from such long-distance trials to rallies, and in July of the same year a determined Healey entered a Super Seven Tourer (not the same car that he had used in the Lands End trial) for the very first of a series of British long-distance events, the Hampshire Automobile Club's Bournemouth rally. The main attraction of this weekend event, held in high summer, was a Concours d'Elegance competition held at the finish—The Autocar dubbed this an: 'open-air Olympia motor show'.

The rally attracted 39 entrants, starting from a variety of far-flung points, such as John O'Groats, Edinburgh, Holyhead and Lands End. It wasn't actually an RAC rally—that event would not be invented until 1932—but the format already sounded familiar. Entrants could choose any number of distant starting points, but were then obliged to call at certain towns for checking purposes. The route between controls was free, but entrants could average no more than 20 mph, as this was still the country's nationwide speed limit.

In the 1920s, rallying success was usually measured by how far, not how fast, a competitor travelled to reach the finish. For every ten miles of direct route covered on the 'Bournemouth' the competitor gained ten marks. Naturally this meant that competitors starting from John O'Groats could amass the highest number of marks—and naturally Donald Healey chose to start from that point. At this juncture we should pause and consider that Healey, and his co-driver, the local Perranporth dentist, Harold Jones, had to drive about 800 miles before they started the event, and only then begin to tackle the 809 mile route down to Bournemouth!

20 mph? Easy? Well, maybe—but consider that the Super Seven Tourer's cruising speed was probably no higher than 45 mph, and that there were no by-passes or motorways in those days. There were no rest halts, which meant that the little Triumph was almost continuously on the road for more than 40 hours.

The climax of the rally was a strict regularity average speed contest, from the final public control at Southampton, to the finish in Bournemouth. One secret check along the specified route, manned by A.V. Ebblewhite (the Brooklands time keeper), caught any errors—and since competitors were asked to be accurate to within 1/10th of one mph, everyone was penalized.

Thirteen cars started from John O'Groats, 11 of which occupied the top 11 places. Right at the top, with a total penalty of a mere 0.2 marks on the regularity run, was the Healey-Jones Triumph Super Seven, thus recording Triumph's first-ever major success in long-distance motorsport. It was the first of many, and the car—registered RL 8125—deserves to be remembered for that reason.

Triumph, naturally, were delighted, especially as their little car had defeated sports cars from Frazer-Nash, limousines from Rolls-Royce, and expensive machinery from Sunbeam, Talbot and Hotchkiss. The winning car, in fact, was not only the winner but one of the smallest *and* the cheapest to compete.

In 1929 the irrepressible Healey then decided to tackle the Monte Carlo rally, persuaded Triumph to lend him a Super Seven saloon, and elected to start from Riga on the shores of the Baltic, where high bonus marks (for distance covered) would be gained. This car, registered WK 7546, was not only equipped with twin spare wheels—each one sticking up high on each side of the bonnet—but had an outsize fuel tank and de-ditching gear. It carried two other passengers, Donald's brother Hugh and a local friend, Lewis (Lew) Pearce. Its top speed was estimated at 45-48 mph.

This was a memorably difficult year, in which only 24 of the original entry of 93 cars succeeded in reaching the finish of the event, and there was to be no fairy-tale debut for the 'works-provided' Triumph. Healey's first problem was actually to get to the start, for he could not proceed beyond Deutsches Krone, on the East Prussian/Polish border, and eventually elected to start from there instead.

Cold-weather preparation and protection in those days, more than 60 years ago, were primitive in the extreme, for the only way of adding grip to the tyres was by using chains (which speedily broke when the snow and ice was left behind). There was no reliable way of defrosting the inside of the screen—and there was no heater.

Even so, the gallant little car kept up to schedule all the way to Paris, and right through to the edge of the Riviera, but from Frejus to Monaco it had to battle with heavy traffic, and eventually reached the final control just two minutes out of time. The only consolation was that Healey then drove it in the Mont des Mules speed hill-climb which followed on the day after the finish—and won its class.

Later in the year Triumph provided two other Super Sevens—one a supercharged two-seater, the other a saloon—for Healey to compete in the Brighton and Barcelona rallies. The Brighton rally, which took place in July, was run on the same format as the 1928 Bournemouth event, with a long road section, and a 50 mile regularity run in to the finish.

As you might expect, Healey, using an open-topped two-seater registered WK 9399, chose to start from John O'Groats, then incurred a total regularity time penalty of only five seconds. It was another victory for the Cornishman, who was rapidly confirming his reputation as Britain's most successful rally driver.

In the Riga-Barcelona rally, which was held in conjunction with the World Fair, Healey used a saloon registered CV 821 (this was a Cornwall number, which indicates that it was locally registered in Perranporth), started from Riga, and completed the 2,000 mile route at an average speed of 25 mph. Running two-up (his third crew member was taken ill just before he left the UK to drive to the start), he won his class, and naturally put up the best performance by a British car.

The spoils of victory—Donald Healey, with his winning Super Seven, after the Brighton Rally of 1929. (Geoffrey Healey)

Donald Healey (left) poses beside the Triumph Super Seven saloon which took him safely and successfully to a class win in the Barcelona rally of 1929. He started from Riga, the same town he had nominated as his start point in the Monte Carlo rally earlier that year. (Geoffrey Healey)

Donald Healey's performance in the Monte Carlo rally of 1930 was nothing short of heroic. His Triumph Super Seven ran three-up throughout the wintry event, lost no time points on the road, and put up the equal best performance in the regularity section at the end of the rally. Because of an unfavourable distance/handicap system he finished seventh overall, not first! The crew (left to right) were: Hugh Healey, Lewis Pearce and Donald Healey. (Geoffrey Healey)

Then, in January 1930, came Triumph's best performance so far in rallying. Using the same car which he had driven from Riga to Barcelona, Healey—along with Hugh Healey and Lew Pearce—started the Monte Carlo rally from Tallinn, in another of the Baltic states, 400 miles further north from Monaco than Riga. On this occasion the weather was much milder than it had been in 1929, for no fewer than 88 competitors (of the 115 starters) reached the finish on the Riviera. Not less than 70 of them gained the maximum possible 'distance/average speed' points.

The Healey/Triumph route, all at a top speed of about 50 mph, was by way of Riga, Königsberg, Berlin, Brussels, Paris and Lyons. However, his start point was marginally closer to Monte Carlo than was Jassy (in Rumania), from which six crews reached Monaco unscathed. Because of the 'distance = bonus points' format of the Monte Carlo rally, this meant that Healey could not possibly finish higher than seventh—and this is precisely where he ended up.

His consolation on this occasion was that he put up the equal best performance on the regularity test at the end of the rally (with the car which won the event outright, having started from Jassy)—which was two circuits of what was known as the Col de Braus loop—and put up much the best performance by a British driver. It was no wonder, therefore, that Healey was then attracted away from Triumph, first of all to drive for Invicta (where he won the Monte Carlo rally in 1931, finished second overall in 1932, and won three consecutive Glacier Cups in the French Alpine Trial!)

In the meantime, private owners had also been proving the worth of the Super Seven. At the banked British track, Brooklands, Victor Horsman developed two single-seater racing specials between 1929 and 1932, one of which used a supercharged Super Seven engine. The engine modifications were apparently all derived by Horsman himself (these included a new camshaft grind, a modified cylinder head, and twin Amal carburettors on the normally-aspirated car), who showed that his tiny cars could lap at up to 80 mph.

On handicap (Brooklands races were usually run on that basis) Horsman had a series of impressive results in short-distance races, with victories in 1929 and 1930, second places in

1931, and third places in 1930 and 1932. Like many other drivers, however, he became frustrated by the way the handicapping system was regulated, and turned to other things.

In long-distance events the Super Seven soon made its mark, notably in North America, where G.A. Woods completed New York–San Francisco–Vancouver virtually non-stop, while the return trip from Vancouver to New York, by way of Cheyenne, Chicago, Toronto and Niagara meant driving 3,538 miles in 8½ days!

In Australia the Super Seven put up similar sturdy performances, for one car won the Sydney Motor Club's three-day reliability trial in 1929, another completed Sydney-Melbourne (575 miles) in a mere 13¼ hours, while a third car set a new Brisbane–Sydney Light Car record, 672 miles in 22 hours 10 minutes.

In 1930 a Triumph won the 1-litre class of the Australian RAC Trial, and was the most consistent car in the RAC-Victoria Double 12 Hour trial, but the most astonishing trip of all was by P.W. Armstrong (a car dealer from Perth), who drove a Super Seven all the way across the continent from Perth to Sydney.

This epic drive—roads in the outback were virtually non-existent at that time—encompassed 3,000 miles, and along with his chosen co-driver George Manley, Armstrong set out to average 18 mph—non-stop! After many tribulations, very little sleep, becoming lost on several occasions and fighting ever-increasing exhaustion, the intrepid pair reached Sydney, by way of Melbourne, in eight days and six hours.

The magnificent little Seven Tourer had averaged 14.9 mph, its only breakdowns involving a change of plugs and a new fan belt. Armstrong, by the way, rubbed in this marvellous achievement by driving the car the whole way back to Perth, averaging 290 miles a day, but stopping to go to bed every night!

1933-1934

In spite of the growing sporting reputation of its cars, however, Triumph seemed to have no immediate urge to set up a formal 'works' competition programme. It is interesting to note that MG, which had no better foundation on which to build sports cars, took the opposite view, and soon built itself a reputation which Triumph might have envied if they had been interested.

Triumph was still mainly a motorcycle business which also produced cars, although car production was expanding all the time, and it was considering a move up-market into the 12hp/14hp 'middle-class' market. Triumph's founder, Siegfried Bettman, was considering retirement, and the favourite to take over management from him was Lt.-Col. Claude Holbrook. Recruited in the 1920s, Holbrook became assistant managing director in 1931, the year in which the company changed its name—to the Triumph Company Ltd—which was an unofficial admission that motorcycles would eventually have to take a back seat.

By 1933 Triumph had changed considerably. Like most car-makers it had suffered badly in the Depression (though there were still some financial reserves to bolster up the balance sheet). Bettman finally stepped down, Lt.-Col. Holbrook became managing director, and the company began fitting Coventry-Climax engines to its new models. Lt.-Col. Holbrook's first major act was to commission the design of a new range of cars, the Gloria range which was due for launch in the autumn of 1933.

Holbrook's problem—which was to be duplicated almost exactly in 1952, when Standard-Triumph faced the same dilemma—was that he wanted to start building a sports car, but his staff had no experience, and no tradition, to back this desire. With a range of Glorias almost ready for launch, he decided to buy in some real expertise. He seems to have had no hesitation in contacting Donald Healey, and summoning him for an interview.

In the three years since he had last driven a Triumph in motorsport, Donald Healey's reputation had gone from strength to strength. His excellent performance in Invictas (which were among the fastest *and* the most tricky-to-handle sports cars of the day) meant that he was now a household name.

After more than two years as a 'works' driver for Invicta, Healey then formed links with Riley of Coventry, not only to drive that company's cars in motor sport, but to work on new sports car designs like the Lynx, MPH and Sprite. Perhaps this explains why it is no coincidence that the MPH/Sprite had some of the same technical and styling 'cues' as later appeared in mid-1930s Triumphs.

There is no doubt that his success in motorsport had made his eventual career aim easier to achieve:

> My successes with Invicta cars seemed to mark the culmination, or very nearly so, of what I had for so long been trying to achieve. This was a close association with the motor industry, so that I could eventually get into it myself—finally to reach my ultimate objective of building a motor car of a type I knew from experience to be what the sportsmen of the day needed.

At the same time he decided to shrug off the strain of keeping up with his garage business in Perranporth, sell it off, and concentrate on a career in the motor industry.

> Donald Healey once told me: The first approach I had from Triumph was from Charles Ridley, who was then Colonel Holbrook's deputy. I was attracted by the Gloria project, which looked better than Riley's own Nine. I also had good friends at Triumph—particularly Gordon Parnell, who was then Chief Designer. After I talked to Charles Ridley, I had an interview with Col. Holbrook.

At this time Healey, although still an employee of Riley, was asked to try out the new Gloria, found that it was disappointing in several respects, and made his opinions clear.

> I told him [Col. Holbrook] that if it was going to compete with the Riley they'd have to get someone on their payroll with sufficient experience to put it right.

Holbrook had apparently already decided that the 'someone' should be Donald Healey, offered him a newly created job as Experimental Manager, and persuaded him to start work almost at once. After stifling qualms of loyalty to the Riley family, Healey moved across town, and started work on the new Gloria models in September 1933.

> My first job was cleaning up the Gloria range, which was just about ready for production. There was no technical director as such. I reported direct to Charles Ridley for a time. . . .

Walter Belgrove, who became Triumph's Chief Body Engineer in 1935, recalled Donald Healey as a real bundle of energy, and summed him up in this way:

> Healey was a great enthusiast. He had the personality to get the best out of a very good engineering team. My direct association with him was a happy one.

This, however, was only the start, because Holbrook also promised him a 'works' motorsport programme, and by 1935 he had been promoted to Technical Director, a post he held until 1939. Although he was always very close to the motorsport effort, he needed a separate competitions manager, so to do this job Healey appointed Jack Ridley, who was Charles Ridley's son and already a very fine competition driver.

In the autumn of 1933 the Gloria range was launched, in several versions, all styled by Triumph's then Chief Body Engineer, Frank Warner. All had an underslung chassis frame, with cruciform cross-bracing, and with half-elliptic leaf springs at front and rear. There was a choice of 1.1-litre four-cylinder or 1.5-litre six-cylinder Coventry-Climax units, built under licence at Triumph, and the performance was kept in check by the use of massive 12 in. diameter drum brakes.

Apart from the four-door saloons, which would make up the majority of sales, there was also an open car called the Gloria Speed Model Tourer, which Healey not only chose as the basis of his first 'works' cars, but also began to modify as a lighter and more sporty car to be called the Gloria Monte Carlo. One of the neat features included in this car's design was the door style, which had a fold-down top panel. With the panel lifted up the door was like

Donald Healey (in the car) and Jack Ridley (his competitions manager) pose in an early example of the Triumph Gloria Monte Carlo, a car Healey specifically developed to make Triumph cars more competitive in 1930s-style rallies.

that of a drop-head coupé, with a high rail, but with it folded down it had cutaway sides to allow the driver more elbow room for sawing at the wheel.

For their first ever Triumph 'works' entry, in the Monte Carlo rally of 1934, Healey and Jack Ridley took every advantage of the regulations. In those days there was no need to qualify with any form of homologation, so the cars could be 'specials'. Accordingly, two very special 1.1-litre Gloria-based cars were produced, which not only had lighter coachwork, but entirely different frames!

The basis of the cars was an old-style Southern Cross frame, which had straight side members instead of the Gloria double-drop type. Not only that, but the cars were equipped with outsize tyres—9.00-16 in. Dunlop ELPs (Extra Low Pressure)—which could run at a pressure of only 10-12 psi. The overall wheel diameter was 32 in. instead of the normal 28 in., but the axle ratio was changed to compensate for this.

Rallying, 1934-style. This is one of the 'works' Triumphs which competed in the Monte Carlo rally of 1934, based on a Southern Cross frame, but with modified Gloria running gear. Donald Healey is in the centre of the shot, with journalist Tommy Wisdom to his left. The third man, who was not competing in this car, is not identified. (Geoffrey Healey)

Smiles of relief, on the streets of Monte Carlo, after the 1934 Monte Carlo rally. The car looks filthy at this point, and standard wheels and tyres have replaced the ELP 'balloons' at this moment.

This meant that the cars had a lot of ground clearance (Donald Healey himself suggested that they looked: 'as if they were on stilts'), that they could float over all manner of rocky terrain, and could also keep going on surfaces where grip was at a premium.

The body style was altered considerably. It was simplified and lightened, though a four-seater layout was retained. For his own crew Donald chose to travel with motoring journalist Tommy Wisdom and racing driver Norman Black (in KV 6905), while Jack Ridley used the sister car KV 6904 and had R.C. Clement-Brooks as his passenger.

As ever, Healey searched for the last little advantage, choosing to start both cars from Athens, which was furthest (no less than 2,352 miles) from Monte Carlo. None of the other British crews in the event chose the Athens start—one reason being that the drive through the Balkans could be extremely wintry, and the roads were often blocked.

Surprisingly, the Athens route was not afflicted by snow, but by mud, for there had been unseasonable heavy rain in the Balkans. No fewer than 15 of the 25 starters came through unscathed from Athens, which made them almost certain to be top contenders for victory. Among them were the two Triumphs, but Healey then had to face up to several much more powerful cars in the acceleration-and-braking driving test on the Promenade at Monaco.

Although the Triumph could not beat the Gas/Trevoux Hotchkiss, nor Chauvierre's Chenard-Walker, it set up an amazing time, only 1.1 seconds slower than the winning car over the 100 metres dash. The result was that the Healey/Gloria combination finished third overall, and won the up-to-1.5-litre category, while Jack Ridley finished sixth in that class. All five of the privately-entered Triumphs also finished, J. Beck's car being best placed at 10th in its class.

Was it any wonder that someone protested against the Triumph, on the grounds that its performance was phenomenal for such a small car, but the engine was measured and found to be in order?

Not only that, but on the complex 'wiggle-woggle' driving test which was organized as a separate competition on the day after the rally ended, Healey set fastest time and won the Coupe de Monte Carlo.

It was a splendid beginning for the team. Healey's standing within the Triumph organization was immediately confirmed, and it needed little persuasion for Col. Holbrook to approve a full-scale assault, in high summer, on the high passes of the Alpine Trial.

For this occasion Healey and Jack Ridley prepared no fewer than six new cars (the money seemed to go a lot further in those days)—all of which were specially produced 1.1-litre cars based on the design of the 'Monte' Glorias, but with different tourer bodies (with two-seater hoods and with the rear seats removed to make space for spares and tools) which Donald Healey has described as four-seaters.

The wily Healey split his resources for this assault, nominating three cars to compete for outright victory in the Coupe des Alpes, and three so-called 'private entries' to go for individual Glacier Cups. The 'works' cars were driven by Col. Holbrook himself, Jack Ridley and Victor Leverett (who had become Triumph's sales manager), while the individual entries were driven by Maurice Newnham (Triumph's London dealer, who would later take the helm at the factory), Miss Joan Richmond, and Donald Healey himself. On that occasion Healey's co-driver was Lewis Pearce.

Then, as in the 1950s and 1960s, the Alpine was a real helter-skelter of an event, taking all the high passes, usually in blisteringly high temperatures, with little time for rest, either for the cars or the crews. Passes (Cols) later to be familiar to TR drivers—Cayolle, Vars, Izoard, Glandon, Stelvio—were all included, but there was the challenge of the dusty and incredibly steep Turracher Hohe in Austria.

The event started from Nice, but passed through France, Italy, Yugoslavia and Austria before finishing in Germany, in the city of Munich. This was the event in which two British teams—one from Triumph, the other from Talbot—put up a series of outstanding performances.

Even on the Stelvio Pass which, at 9,000 feet was then the highest through road in Europe, the Triumphs were required to average 22 mph/35 kph for the 49-hairpin, 12-mile ascent. Only one car—that driven by Joan Richmond and Mrs Gordon Simpson—failed to meet the target of a completely clean run throughout the rally, and then only because her car slid off into a rock face at one of the hairpin bends and broke a front spring. Later she had to mend the spring, and was disqualified for not using tools which she had been carrying on board since the start. She had borrowed a sledgehammer and an anvil at the other side of the Stelvio pass!

Donald Healey poses proudly in front of the much-modified Triumph with which he had just finished the 1934 Monte Carlo rally. His competitions manager, Jack Ridley, is alongside the second of the Triumphs. Healey himself finished third overall, and won his class, while Ridley finished sixth in that class. (Geoffrey Healey)

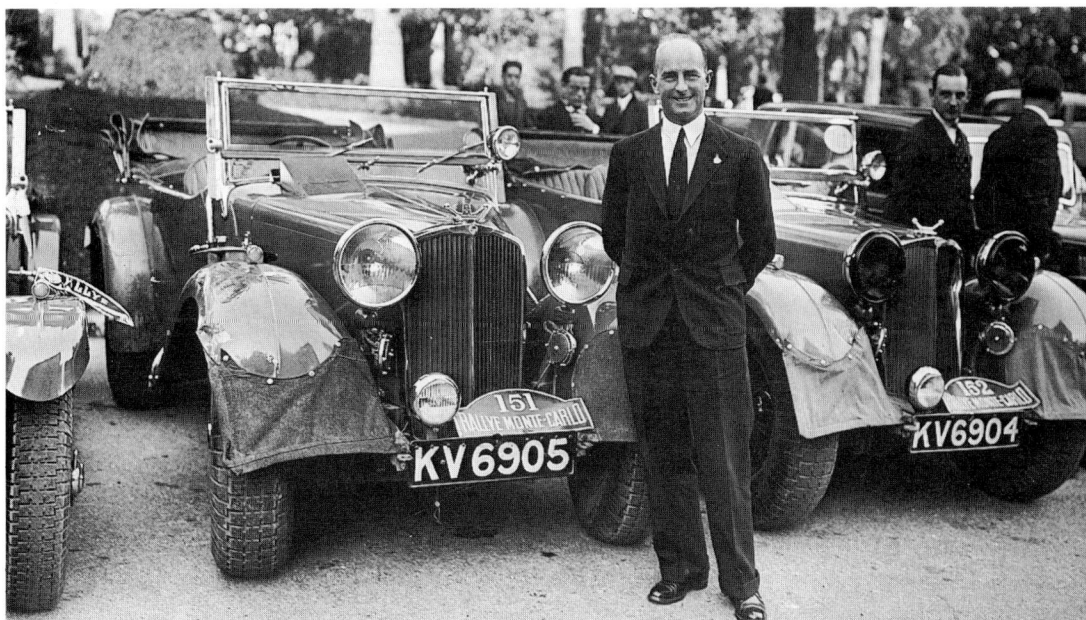

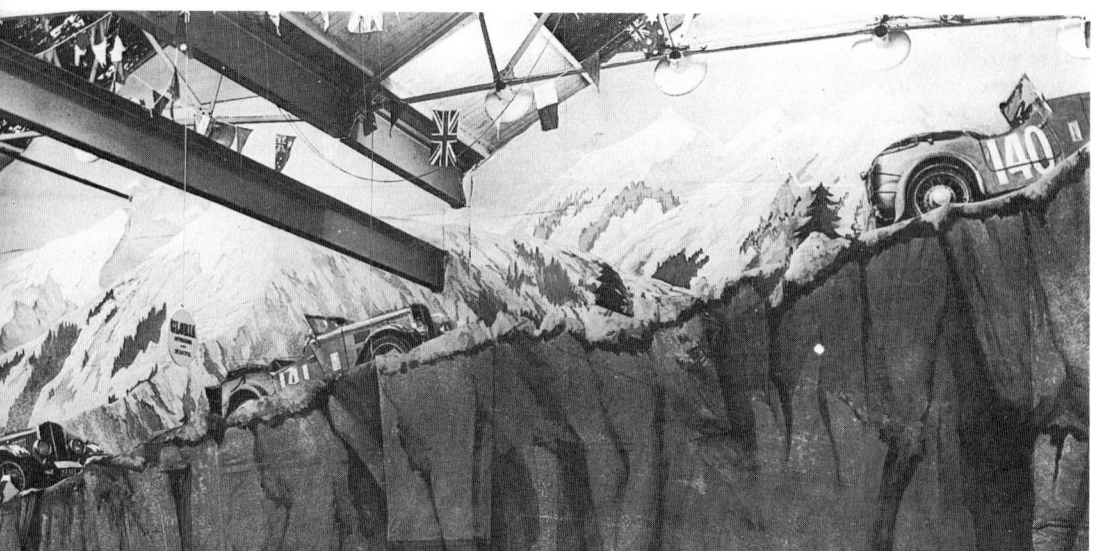

Triumph was so pleased by the performance of its 'works' cars in the Alpine Trial of 1934, that it arranged this startling display of the 'works' cars inside the Triumph factory, to coincide with a trade occasion, probably the launch of the 1935 model range.

Nevertheless, it was another triumph for Triumph. The official team won a *Coupes des Alpes* in the team contest, while Donald Healey and Maurice Newnham won individual *Coupes des Glaciers*. Triumph's performance was by far the best in the event, for the team finished well ahead of the opposition from Singer and Adler-Trumf (of Germany).

But even this Alpine rally success was not the most exciting event in Triumph's motor-sport history in 1934. An entirely new supercharged car was about to be unveiled.

The Dolomite Straight Eight project

Every Triumph enthusiast, surely, knows that the magnificent Dolomite straight-eight sports car was launched in 1934, that it was very closely based on the layout of the Alfa Romeo Monza sports car, and that to make this possible Donald Healey had 'done a deal' with Alfa Romeo's technical chief, Vittorio Jano.

Everyone, too, knows that only three cars were built, that Donald Healey himself was spectacularly eliminated from the 1935 Monte Carlo when his Dolomite was hit by a train, and that the project itself was killed off in 1936. Everything to do with this project was then sold off to Tony Rolt, who raced a car for a time, after which the cars disappeared into limbo for many years.

In the 1980s not one, but both the surviving cars, rebodied and much modified, reappeared, were restored, were sold to wealthy collectors, and have now become well-known elements of Triumph's classic folklore.

The saga surrounding the birth of the Dolomite straight-eight has now been told in several other books, so I will not repeat it here. In summary, though, let's just point out that once Bentley had withdrawn from motor racing, and Alfa Romeo's supercharged straight-eight models had come to rule the roost, various British enthusiasts wished that it could be beaten.

Sir Henry ('Tim') Birkin tried to raise the finance for such a project, and so did Earl Howe, but it was a meeting between Col. Holbrook, Donald Healey and journalist Tommy Wisdom which saw Holbrook agree to the design of a new Dolomite, which had entirely new chassis, body style, transmission, and suspension components, along with a direct copy of Alfa Romeo's famous supercharged twin-cam eight-cylinder engine.

Healey, it is said, wanted to produce a competition car to beat the sports car world at its own game, and in spite of being granted a derisory budget to develop the new model—Donald Healey once told me that this was only £5,000, which was nevertheless enough to buy two Rolls-Royce Phantom II limousines in those days—his team produced enough parts to build three prototypes.

The lasting mystery behind this project is not the connection between Triumph and Alfa-Romeo, but why the Triumph was built in the first place. In its current parlous financial condition Triumph could surely never have afforded to put it on sale—and the rules of motor racing were just changing to make the use of supercharged cars uncompetitive!

Nevertheless, the first car was completed in 1934, and when *The Autocar* tested ADU 4, it recorded a two-way top speed of 102.47 mph, which made it the fastest British road car of the period. If only it could be put on sale, and if only it proved to be reliable, it might have a great future.

1935 and 1936

After the Dolomite Straight-Eight was previewed at the Olympia Motor Show, Healey and Jack Ridley laid their plans for the next Monte Carlo rally. Healey, who wanted to repeat his 1931 success in the Invicta, decided to prepare a Straight-Eight (his co-driver was to be Lew Pearce). At the same time Jack Ridley and Vic Leverett were entered in two-seater sports-bodied versions of the 1.2-litre Gloria, actually two of the cars used in the Alpine rally of 1934.

On this occasion there was no 'mileage advantage' to be gained by starting from one or another point, as the event was to be settled on the driving test at the end of the event, so all three crews were sent to begin their Monte from Umea in Sweden. The Dolomite had been most carefully prepared for the event, with the best soft-top weather protection that could be devised, with oversize wheels, with two spare wheels, with hard multi-leaf rear springs, and with an extra movable spotlamp mounted on the windscreen pillar.

ADU 4 was the Dolomite Straight-Eight with which Donald Healey was to compete in the 1935 Monte Carlo rally. Here it is seen, in high-speed testing, at Brooklands, during 1934. The headlamps have been turned sideways to decrease their drag. The passenger is Jack Ridley, but the test driver (not Healey himself) is not identifiable.

ADU 4—Donald Healey's Dolomite Straight-Eight, all prepared for the Monte Carlo rally of 1935, poses at Brooklands before leaving for the start. Lewis Pearce (left), who was to be the co-driver, is clearly not yet kitted-up for the journey.

Although everything started well, while he was picking his way carefully through Scandinavia in the fog and inky darkness of a January night, Healey arrived at an un-gated railway crossing at exactly the same time as a train:

> As we approached the Danish frontier I was chatting with Lewis, following another car in the darkness to save our batteries, when suddenly and terrifyingly we heard a shrill, almighty scream. I looked at Lewis: 'The ruddy supercharger's seizing!' I said. I'd scarcely completed the words when there was a shattering crash, and we span round in a complete circle in the middle of the road. . . .

Donald Healey, as happy and energetic as ever, alongside his Dolomite Straight-Eight before the start of the 1935 Monte Carlo. This shot was probably taken at Umea, in Sweden. (Geoffrey Healey)

This page: Stories without words! This sort of winter seemed to be normal for Scandinavia in the 1930s. Donald Healey's Straight-Eight Dolomite was running in convoy with Jack Ridley's ex-Alpine Trial two-seater Gloria 'somewhere along the way'. Note the twin spare wheels, on the Dolomite.

> Then we heard the sound of escaping steam and, looking up from beneath the edge of the low hood, we saw the lights of a train. . . . It had been a train's whistle, and it had hit our right-side, front corner on an unguarded level crossing.

The car was badly damaged, and the crew were lucky to be alive, so there was nothing more to be done except arrange to get the wreckage back to the UK. On its return, Col. Holbrook was sympathetic to the crew, and agreed to it being rebuilt for another occasion.

But it was not all bad news for Triumph, for Jack Ridley fought his way through to the Riviera, had a supercharger fitted just before he arrived at Monte Carlo (the regulations did not ban this sort of en-route modification), then set up a stunning performance in the wiggle-woggle test. As *The Autocar*'s reporter wrote:

> The little black two-seater was magnificently handled from start to finish—and streaking through the complicated turns and reverses without a mistake put up a splendid time.

The result was that Ridley (driving KV 6905) finished second overall—one place higher than Donald Healey had achieved in 1934—and he also won the entire 1,500 cc class. The only car which beat him was Charles Lahaye's big Renault. There had been 149 starters and 102 finishers.

In 1936 the indomitable Healey entered a Dolomite in the Monte Carlo rally. This time (as in 1930) he chose to start from Tallinn—on the east side rather than the west side of the Baltic Sea.

Although ADU4 had been comprehensively wrecked in the 1935 event, it was miraculously reborn for 1936, looking exactly the same as before except for the different positioning of the rally plate at the rear. This time, however, to make it more suitable for the rally (where torque was more important than peak power), Healey built another version of the engine; the engine was enlarged, but the supercharger was removed.

The size of that engine has been variously quoted as 2.4-litres (Donald Healey, 1973) or 2.6-litres (Donald Healey, 1988)—but when one of the cars came to be restored in the 1980s it was found to have a 67 mm cylinder bore and a 2.5-litre capacity! Let's leave it at that.

On that occasion Healey battled through from Tallinn, unpenalized on the road, but could not quite match the times set up by special cars with special brakes on the 'wiggle-woggle' test in Monaco itself, and finished eighth overall. Naturally his was the best performance by a British driver, and yet again he had shown that the Triumph-Healey combination was the standard other British makers had to beat.

That, however, was the last we saw of the fabulous 'works' straight-eight Dolomite, for soon after this the entire project—cars, spare engines and all other parts, being sold off to Tony Rolt, who built up one complete car and raced it for a time before he graduated further to a single-seater ERA. Just to confuse everything, however, the name 'Dolomite' came to be applied to a new range of Triumph family cars, which made their debut in 1936!

This was the same Monte Carlo rally in which Joan Richmond and G.S. Brooks used the

In 1936 Donald Healey used a rebuilt Dolomite Straight-Eight—ADU 4 once again—for a second assault on the Monte Carlo rally. On this occasion the car was unsupercharged, and used an enlarged engine. Healey finished eighth overall, and set the best performance by a British driver. (Geoffrey Healey)

On the 1935 Monte Carlo rally, Donald Healey's Dolomite Straight-Eight was destroyed in a collision with a train on a level-crossing which had no barriers. (Geoffrey Healey)

'old faithful' Gloria (KV 6904) to finish second equal in the 1.5-litre class. Because Ms Richmond had a male co-driver she was not eligible for the prestigious Ladies' Crew award which, on her performance in this event, she would easily have won.

By this time, however, Triumph's corporate finances were in disarray, and there was little

Donald Healey also found time to meet the customers, and the personalities who bought Triumph cars. The machine in question is a six-cylinder Vitesse, the character in the middle of the group is air-ace and record breaker Flt. Lt. Tommy Rose, and on his right is Maurice Newnham, Triumph's London distributor. The year is 1936. (Geoffrey Healey)

money available for spending on motorsport. The motorcycle business had been sold off in 1935, a new car assembly factory had been purchased in another part of Coventry, and Donald Healey's duties as technical director were taking up more and more of his time. Among many projects, he had to finalize a completely new range of engines for use in the forthcoming middle-class Dolomite models.

Although he found time to compete in the Alpine Trial of 1936, where he used a six-cylinder Vitesse saloon (which used the new Dolomite-type of overhead-valve engine), and won an individual Glacier Cup (his second), his last major entry in a pre-war international rally came a few months later.

1937 and 1938

Even though there had been great controversy after the 1936 Monte Carlo rally (where all manner of outlandish 'special' cars had been allowed to compete), entries for the 1937 event actually increased, and 121 cars started. One of those entrants, by the way, was a young man called Maurice Gatsonides, who would become a very important 'character' in the Triumph team in the 1950s.

Perhaps the rise in entries was because the organizers had decided to drop the infamous 'wiggle-woggle' test, and sort out a result on a 100 km strict regularity test on the final run down to Monte Carlo itself, allied to a simple acceleration and braking test on the Promenade in Monaco.

Even though there was still no restriction on the type of cars which could be entered—there was still no need to use standard production cars—the single 'works' Triumph was basically standard, and was a nicely-detailed two-door Gloria saloon, with a 1496 cc four-cylinder engine, which was to be driven by Donald Healey, Tommy Wisdom and Norman Black.

For his last 1930s attempt on the Monte Carlo rally, Donald Healey used a much-modified Gloria saloon, and elected to start from Palermo, in Sicily. The crew on this occasion were (left to right): Donald Healey, Tommy Wisdom and racing-driver Norman Black. (Geoffrey Healey)

The 1937 'works' Gloria saloon, covered with snow (which means that there can have been no in-car heating to melt it!) at the Ljubljana control of the 1937 Monte Carlo rally. As usual, Donald Healey seemed to be enjoying himself—but a few hours later he was to go off the road. (Geoffrey Healey)

As *The Autocar* pointed out in its pre-rally comments, this was a very thoughtfully-equipped car:

> Donald Healey's Gloria short-chassis saloon is of another type. The normal wings and running boards have been removed, and cycle-type wings fitted, because with them it is easier to unditch the car, and the undershield has been made quite smooth for the same reason. There are night

Donald Healey, one presumes, took this photograph of his own 1937 Monte Carlo rally car, for that is Norman Black (left) and Tommy Wisdom (right) alongside the car at a refuelling halt. (Geoffrey Healey)

lights in cases to provide heat to act as a defroster for the windscreen, and fuel pipe lines are duplicated, the petrol tank armoured, and, besides chains, the car carries the Swedish Autotraktor ratchet unditching gear. But perhaps the most exciting part of the equipment is that the navigation instruments include a rotating chart giving times for minute fractions of the 100 km Regularity Test, and a special recording clock with one hand revolving once in 72 sec. to suit this test.'

Once again, there was no points advantage in starting from a particular place, so this time Donald Healey chose to start from Palermo in Sicily, not merely because the winter weather through Italy was likely to be better than from other starting points, but because this would give him a chance to look at the regularity section—in other words, to do a recce—on his way down to the start. No fewer than 28 crews had the same idea!

Palermo starters drove by way of Rome, Ljubljana, Vienna, Strasbourg and Dijon before tackling the regularity run, but unhappily the Triumph did not feature in this. Although Italian weather was fine, there was mud, water, ice and lots of snow in the Balkans. As in previous years, there was really no heating for the inside of the car, and it must have been a trip to purgatory.

Along the way a combination of snow drifts and sheet ice saw the Triumph slide irretrievably off the road, and out of the event, long before the complex new navigational system could be tested. That was almost the end of Donald Healey's 'works' rallying career at Triumph. As he once wrote:

> As technical director I was becoming too busy, too involved with the daily work at the factory to spare much time. Anyway, by now our cars were appearing in a formidable number of events, driven by private owners.

He and his colleague Lew Pearce, however, usually managed to enter Britain's most prestigious event, the RAC rally, though it was hardly tough enough to interest a man who had won Monte Carlo rallies and Glacier Cups! In 1937 and again in 1938 Pearce won his class in an event which involved no more than a long drive to a seaside resort, where complicated driving tests sorted out the result.

From 1938 Triumph confined itself to rallying cars like this Dolomite Roadster in British events. These cars were bulky, but fast, and were regular class and category winners.

In 1937 he drove to Hastings, to win Group 4—the 10 hp to 15 hp class for closed cars, using a 2-litre Dolomite-engined Vitesse saloon—while in 1938 he drove to Blackpool to repeat the trick in a 2-litre Dolomite saloon. In 1939 a Triumph Dolomite completed its hat-trick in this class, this time with a privately-entered car driven by G.S. Davison.

That, however, was the end of Triumph's motorsport effort. By 1939 the company was in desperate financial trouble, and from June it went into receivership. There would be no more 'works' Triumphs until the 1950s.

Chapter 2

Jabbeke, Alpine Rally and Le Mans

The Ken Richardson era begins, 1953-1957

After Triumph's bankers called in the receiver in June 1939 the company rapidly went into hibernation, and Donald Healey soon drifted away, eventually to found his own company. It would be 14 years later, after two changes of ownership, before there was any renewal of Triumph motorsport activity.

On the very eve of the Second World War Triumph was taken over by Thos. W. Ward Ltd of Sheffield, but the factories were badly bombed in the blitz of 1940, and the remnants of the business were sold off to the Standard Motor Co. in 1944.

Standard had come a long way since the late 1920s, when Capt. John Black joined the company as its general manager. By 1934 he had become Standard's managing director, and by the end of the decade he was its despotic dictator. Standard had become one of Britain's 'Big Six' car makers.

Even so, there was no sporting instinct in the company. The design team was conventional to a fault, and the cars themselves were uninspiring. Although Raymond Mays chose Flying Standard running gear as the basis for his own road-car project in 1939, there was no sporting connection of any sort, and no question of a 'works' motorsport operation.

Towards the end of the war, John Black had been knighted, and his personal assistant was a young man called Alick Dick. Dick, who later went on to become Standard's managing director in the 1950s, once told me:

> Just about the end of the war, Triumph itself was bust. John Black wanted another make name beside Standard—just that. . . . We looked at the place, which wasn't worth a farthing; we bought the firm . . . we sold the factory to the B.O. Morris company, and all we got was the name.

Were Standard interested in the pre-war designs?

> Not one tiny bit. We never even gave it a thought. But what you have to remember is that before the war Standard had been supplying parts to SS. Standard had special engine tooling for the six-cylinder engines, and the overhead valve version of the four-cylinder engine, of which Lyons took all supplies. Immediately after the war Bill Lyons wanted to buy the tooling—he wanted to make all of his Jaguars in future—and John Black was willing to let him have that so that we could build a competitive car. We kept the four-cylinder tooling, but it just wasn't viable without a new chassis and a new name. So that's why we bought Triumph. Just because Bill Lyons made a sporting saloon, or a sports car, John Black was not going to let him get away with it.

This confirms that there was no question of reviving old models, for these were not only old-fashioned by 1945, but all the tooling had been destroyed in the blitz. There was one significant personal link between the new and the old, for the 'old' Triumph's chief body engineer, Walter Belgrove, had joined Standard in a similar capacity, and effectively carried out the same

The works TR story really started here! The famous 'wet-liner' engine was originally designed by Standard-Triumph to be used in two immediate post-war products—the Ferguson tractor, and the Standard Vanguard. At this time there was no thought of turning the engine into a competitions unit.

job until the mid-1950s. New models would have to be based on Standard bits and pieces, which means that all post-war Triumphs were more properly described as Standard-Triumphs.

This also explains why I have chosen to detail the exploits of Standard-badged cars in this period as well as those of Triumph, for the two cars became ever closer to each other. In many cases they used common components, in all cases they were designed by the same team of engineers, they were always assembled alongside each other in Coventry—and the 'works' cars were always controlled by the same management.

Sir John's strategy for the rebirth of Triumph could best be described as confused, for until the 1950s (after he had left the company) there was no lasting and consistent use of the new marque. In no case was there any attempt to use the traditions or the reputation of the past—but merely to use 'Triumph' as an up-market alternative marque to 'Standard'.

The original post-war 'Standard-Triumphs' were an intriguing mixture, for they used Standard-built engines which had originally been developed for use in the SS-Jaguar 1½-litre; transmissions, axles and suspensions from the 1930s-style Flying Standard saloons; tubular chassis frames, and aluminium panelled bodies.

Later Sir John inspired the development of the stubby little Triumph Mayflower saloon, allowed Walter Belgrove to produce the bulbous TRX Triumph Roadster of 1950, then made an abortive attempt to take over the Morgan sports car company in 1951. His decision to have a new sports car designed—which became the TR2—followed early in 1952.

Standard seemed to be floundering in its approach to Triumph, this being compounded by the calibre of its design staff. Although young men like Harry Webster and Lewis Dawtrey were keen enough, and interested in cars and sporting motoring, their technical director, Ted Grinham, was not. Grinham was a long-serving Standard employee, a dour individual who had no experience of such cars. Worse, there was no sporting element in the development team either.

Accordingly, throughout this time Standard (and therefore Triumph) seemed to have no interest in motorsport—except, that is, that Sir John Black agreed to back the calamitous BRM Grand Prix car project with money, expertise and facilities. I cover this in more detail in Appendix 1 at the end of the book.

This, then, is the point at which Ken Richardson, Standard-Triumph's competitions manager between 1954 and 1961, comes in to the story, for before joining Standard-Triumph he had worked with Raymond Mays at ERA and BRM. In 1952, when the prototype TR2 was completed, he was between jobs.

Born in Bourne, Lincolnshire in 1911, and later apprenticed to an engineering concern, Richardson joined Raymond May's English Racing Automobiles (ERA) team in 1934, worked for Rolls-Royce on aero engine development during the Second World War, then rejoined Raymond Mays, at BRM, in 1946.

Before long he began test driving Mays's cars, but his first actual race was in the legendary Ferrari-based 'Thinwall Special', in the British GP of 1949, where he unfortunately crashed. Later, he did much of the test driving of the vee-16 BRM, and in 1951 was nominated to drive the car at the Italian GP, until the RAC refused him permission to start on the dubious grounds that he lacked experience; in practice he had actually been lapping faster than the other 'experienced' driver, Reg Parnell.

At this time there was regular turmoil in the British Motor Racing Research Trust (which controlled the BRM project), and one result, as Raymond Mays later wrote in BRM (a book long out of print), was that:

> The Trust also decided to dispense with the services of my old colleague and mechanic Ken Richardson, leaving us without a resident test driver. I was sorry to see him go, for our racing association had been long and friendly.

For Ken, however, this was the darkest hour before the dawn, for in Coventry in the summer of 1952, Standard-Triumph was rushing to complete the prototype of its new 2-litre sports car, coded 20TS, but they had absolutely no expertise to develop it.

Quite out of the blue, he was approached by Ted Grinham just two weeks before the new car was to make its debut at the London Motor Show. As Ken recalls, Grinham said to him:

> Nobody at the Standard Motor Co. Ltd knows anything about sports cars, being quite frank. Would you like to join us?

Even though Richardson himself was no expert on the subject (for then, as later, there was a world of difference between racing cars and sports cars) he thought he could do the job, and accepted the post of Development Engineer, Triumph Sports Cars. However, as he once recalled in a *Motor* reminiscence in 1973:

> I did not see the bob-tailed TR1 which I was supposed to develop until I examined it on the Triumph stand at Earls Court during the 1952 London Motor Show. When I saw the ruddy thing I thought: 'Oh my God, what have I let myself in for?'

Like many other people, Ken uses the unofficial nickname of 'TR1' for a car which *never* officially carried a name—there has *never* been a TR1 at Triumph—but his contribution to the development of that ungainly prototype is now well known, and acknowledged. That story has been told several times before, and it proves that grit, sheer hard work and—in Richardson's case—the bloody-minded determination to 'get it right' can work wonders.

Having tested the original car at Sir John Black's invitation, and described it as a 'death trap', he was hired to help to put it right. Along with Harry Webster, who was Standard-Triumph's chief chassis design engineer, and a small team which included Lewis Dawtrey, Jim Parkinson and John Turnbull, he transformed the original poor design into a service-able sports car in a matter of months.

Then came the first occasion which pushed Standard-Triumph towards a motor sport programme. Over on the other side of Coventry the Rootes Group had produced a smart, though heavy, new sports tourer. They called it the Sunbeam Alpine, and sent Sheila Van Damm and Stirling Moss (two of their 'works' rally drivers) to Belgium, where on 17 March 1953 Stirling achieved a two-way average speed of 120.459 mph, and Sheila achieved 120.135 mph over the flying kilometre on the Jabbeke highway.

Soon after this Sir John Black called Richardson, asked him if the new TR2 could do at least as well, and told him to achieve it! This, no question, was the start of the modern Triumph 'works' competition programme, and Richardson had only eight weeks to prepare a car. Unlike Rootes, he couldn't hire a star driver, so decided to drive the car himself.

By careful preparation of the engine (which included reversing the recommendation of a so-called engine 'expert' in the Engineering Department), and by developing special streamlining for a prototype car, and making repeated high-speed tests on the 'Bicester straight' near Oxford, Richardson was ready within two months.

Because Standard-Triumph intended to use a near standard engine, a higher top speed had to be achieved by improving the streamlining of the original car. To do this, the front and rear bumpers were removed, the windscreen was replaced by a small plastic wind deflec-tor ahead of the driver, wheel spats were added at the rear, a rigid 'tonneau cover' was added over the cockpit, and a full length undershield was also fitted. Dunlop Road Speed tyres were used.

On 20 May 1953 Richardson himself drove the car which had been prepared in this way—the second prototype TR2, MVC 575, which had only been registered in March 1953—in an attempt to beat all the Sunbeam's marks, on the same piece of dual carriage-way motor road in Belgium. This, the Jabbeke highway, was a recently completed section of dual carriageway motorway which ultimately would link Brussels to the coast at Ostend. It was dead flat and for the time being it was possible to have it closed to other traffic when car makers were in search of really sensational marks.

Early in the morning, when the road was closed off by the police, it was dry but misty, but by 06.30 the skies had cleared and the wind speeds were negligible. All the observers—who

The prototype TR2 at speed on the Jabbeke road, in Belgium, in 1953. In speed trim, with an aero screen and tonneau cover fitted, the car achieved 124.889 mph. With the windscreen in place, and the hood erect (as in this shot) it ran at 114.89 mph. (BMIHT/Rover Group)

included Sir John Black, Ted Grinham, and several members of the motoring Press—stood at one end of the measured distance, and were perturbed to see the light-grey car flash by with the engine sounding distinctly rough.

When Richardson reported back to the control, it was to tell them that he had achieved 104.86 mph on three cylinders—for a plug lead had become detached at the start of the timed run. Not bad for a 3-cylinder 1½-litre sports car—but not what was intended!

In the next series of runs, in the streamlined, 'semi-single-seater' condition, the car was a lot quicker than before, and after the mechanics had virtually restored the car to its standard condition (though with the full length undershield still in place), two more sets of excellent speeds were recorded. These were the figures which Standard-Triumph flashed all round the world in the next few hours:

Distance	Mean (two-way) speed
In speed trim, with metal tonneau, undershield, no screen, using overdrive	
Flying kilometre	124.889 mph
Flying mile	124.095 mph
In touring trim, with windscreen fitted, with hood and sidescreens erect, but with wheel spats and undershield retained	
Using overdrive:	
Flying kilometre	114.890 mph
Flying mile	114.213 mph
Not using overdrive:	
Flying kilometre	108.959 mph
Flying mile	108.499 mph.

What is truly interesting is that this car was obviously using a standard engine—and the inference is that the full length undershield and the wheel spats only did a little to improve the penetration and the performance!

The reason I can say this is that when the TR2 production car was road tested by independent magazines in 1954, the test car—an overdrive-equipped OHP 771—was almost as fast. *The Autocar* recorded a two-way speed of 103.5 mph in December 1953, while *The Motor* took the same car to and from the Geneva Show a few weeks later and achieved 107.3 mph, no doubt because the car had loosened off a little in the interim period.

Sir John Black was delighted by the Jabbeke performance—not only had his precious TR2 beaten the Sunbeam Alpine's figures, but it had also beaten those set by a prototype Austin-Healey in the autumn of 1952. Ken Richardson was delighted because it meant that he could get back to completing the development of the new car.

TR2 production then began, haltingly, in the summer of 1953, but by the end of the year only 248 such cars had been delivered. Standard-Triumph still had no declared motorsport policy for the car—yet private owners immediately started to use their cars in races and rallies.

1954

At this point a buccaneering motor trader from Manchester, Johnny Wallwork, enters the story. His garage had picked up Standard and Triumph agencies in 1937, and he had started his post-war rallying, as a privately-financed hobby, in a Standard Flying 12, into which he inserted a Flying 14 engine.

Next he got interested in production car trials, bought himself another Flying Standard chassis, shortened the wheelbase by 18 inches, put a little aluminium body on it, and started to enjoy himself. At the same time a motor trader friend of his, Ken Rawlings of Birmingham, also built himself a Standard-based special which he painted yellow and called 'Buttercup'.

[Incidentally, some enthusiasts say that Sir John Black was inspired to produce the TR2 after he had taken a good look at these cars, especially 'Buttercup', and Johnny Wallwork insists that his 'special' made a trip to Coventry for it to be inspected. On the other hand, neither Harry Webster nor Alick Dick, when questioned, admitted to having seen either car, at any time!]

During 1953 Wallwork spent some time recovering from a bad rallying accident in a 2½-litre Lea-Francis sports car, which left him with a big scar on his face—the inspiration for his subsequent beard. By the end of the year, however, he was back in harness, itching to return to rallying:

> I talked to Lyndon Mills, and persuaded him to let me have four of the very first TR2s, for friends, all competition drivers, that I had a high regard for. They went to Peter Cooper, to Bill Bleakley from Bolton, and to Mary Walker from Scotland.
>
> We all entered them for the RAC rally, which was to be held in March, and they were so new that there really wasn't time to prepare them, they were almost straight off the production line. My car was barely run in—I'd only had it for two weeks when the rally started—so my experience of the car was still very limited.

The RAC International rally of March 1954 was the first major event to be tackled by the TR2s and, like all good fairy stories, it ended in glorious success for the new car. Not only did Johnny Wallwork win the event outright, but Peter Cooper took second place, Bill Bleakley finished fifth, and Mary Walker took the Ladies' Award! All the cars were privately owned, so the effect on influential motoring opinion was amazing.

In those days Britain's premier event was not the flat-out speed event that it subsequently became, and there was a lot of boring main road motoring to be completed. In 1954 the event started from two locations—Hastings and Blackpool—on Tuesday, 9 March, but the finish, back in Blackpool, was not until Saturday, 13 March. On the other hand, the 2,000 mile route included difficult navigation sections in Wales, Derbyshire and the Lake District, as well as a series of high-speed manoeuvring tests.

Sounds easy? Maybe, but the result was that only 21 of the 240 cars were unpenalized on the road sections. The results were an emphatic triumph for Triumph, as this summary of the leader board confirms:

RAC International Rally 1954

Position	Car (Crew)	Marks Lost
1.	Triumph TR2 (J.Wallwork/J.H.Brooks)	416.67
2.	Triumph TR2 (P.G.Cooper/O.L.Leighton)	435.05
3.	Ford Zephyr (T.C.Harrison/E.Harrison)	440.50
4.	Sunbeam-Talbot 90 (P.Harper/D.R.Humphrey)	441.00
5.	Triumph TR2 (W.D.Bleakley/P.Glaister)	445.85
6.	Alvis 3-litre (R.Adams/L.R.Rawlinson)	449.70

Most significantly for Wallwork, his TR2 was unpenalized on the Silverstone and Prescott tests, in the Welsh and Peak District navigation sections, and on all of the road sections.

At a stroke, it seemed, the TR2 had arrived on the sporting scene, for no fewer than 14 TRs had started in the rally, and the victory had come against dedicated 'works' drivers such as 'Cuth' Harrison, Peter Harper, Ronnie Adams, and Jimmy Ray (whose Morgan Plus 4 had finished seventh).

Only a couple of weeks later Gregor Grant, the extrovert Editor of *Autosport*, borrowed one of Standard-Triumph's TR2 Press cars (OHP 676) and, together with Peter Reece, drove it on the French Rallye Lyon-Charbonnières. Grant, who was more of a *bon viveur* than a rally driver, allowed Reece to do much of the driving, but everyone was mightily impressed to see this virtually standard car finish third in its class (behind a 2.3-litre Salmson and a 3.5-litre Delahaye), and end up as one of only 16 cars to be unpenalized on the wintry road sections in the Alps.

As Gregor later wrote about the car:

> I have nothing but praise for the Triumph. Whilst very expensive machines fell by the wayside, the TR2 never showed the least sign of giving trouble. The engine was as healthy at the finish as when we set out from England. . . . We experienced a certain amount of brake fade, but quite honestly, the anchors stood up to the dicing on the worst sections without the least sign of 'wooden pedal'. . .

A few weeks later, with Stan Asbury as his co-driver, Gregor used the self-same car to take 17th place overall in the Dutch Tulip rally.

Even so, Standard-Triumph had taken no more than polite interest in the running of these events, for at that time Ken Richardson was not involved in any plan to take Triumph back into motorsport, though he has told different stories about his attitude at different times.

In 1973, when recalling his career for *Motor*, he wrote that:

> This [RAC Rally] success made everyone at the works competition-minded and they decided to start a competitions department—much to my horror, for I'd had 18 years in motoring sport, and really wanted to get out of it.

In 1987, however, when interviewed, he was asked if he had any thoughts about competition when developing the TR2: 'Of course, what did you expect with my background?' Talking to technical journalists who watched the project during this time it was quite obvious that they viewed the project with scepticism. Competition has always been the best way of creating consumer awareness for the motor industry.

The germ of a motorsport programme, however, had already been planted long before this, as another important character in the Triumph motorsport story makes clear. That character was an entrepreneurial and persuasive Dutchman, Maurice Gatsonides.

By 1954 'Gatso', as he was known to rallying enthusiasts all round the world, was already a legendary figure in rallying before Standard-Triumph management came to know him. He had tackled his first Monte Carlo and Liege-Rome-Liege rallies before the Second World

War, had startled everyone by taking second place in the 1950 Monte Carlo rally driving a gargantuan Humber Super Snipe, and had then put in an immaculately-rehearsed performance to win the Monte Carlo rally of 1953 in a Ford Zephyr.

He told me:

Immediately after winning the Monte Carlo I was approached by someone from Triumph . . . but for 1953 I was committed to Ford. Maybe I would be free for 1954.

Then, during the 1953 Motor Show, in London, they asked me again. I was asked to see Ted Grinham. First of all I was met by Ken Richardson at Birmingham Airport. Then I went into a meeting with Ted Grinham—just him and me, Ken Richardson was asked to stay outside at this stage—when he asked me if I was willing to work for Standard-Triumph?

He said that he wanted more publicity for the TR2. He wanted to enter cars for the Alpine rally, which was very popular with British manufacturers at that time. But they didn't know whether the car was alright. he asked me if I could advise on preparation, and test the car?

They asked me if I should enter a car in the Tulip rally to test for the Alpine? I said: 'No, because if something breaks in the Tulip rally there will be bad publicity.' I suggested that we should have a very good and heavy test in the Mille Miglia instead.

This was a challenge for myself and for the car. We agreed that if something went wrong in the Mille Miglia we should just retire, and no-one would know why. I also said: 'Let me have one car, completely standard, which I can use for normal work. Then we can build a special car, with all possible modifications, for the Mille Miglia itself.'

This was agreed. Ken Richardson was then asked to come in. Grinham asked him to prepare a car for the race, as fast as possible. He could tune it up, but it must be reliable for 1,000 miles, plus the journey to and from Brescia (from England).

Ken then asked me how we would share the work on the Mille Miglia? Should we share the driving? I said that I wanted the Mille Miglia to be a single-handed driver and, although it might have been a bit of a white lie, I told him I was very much afraid of someone else driving, so that was the only way I would accept this job. In the end I drove every yard, and every minute, of the 13 hours 56 minutes it took us to complete the Mille Miglia.

The agreement Gatsonides made with Standard-Triumph was much more comprehensive than enthusiasts realized, for he was also hired as the company's European Technical Service Representative, which was really a half-time job involving a lot of travel, all round Europe, but which left him free to go rallying for some of the year.

He was actually to work for the factory for more than five years, driving in rallies, liaising with American servicemen in Europe who bought TRs preparatory to shipping them home, carrying out promotional tours in many countries (including Nigeria and Kenya), giving lectures, and even visiting North America to take part in the Great American Mountain Rally:

It paid well, and in the beginning I think Triumph needed the publicity. In 1955, which was my busiest year, I travelled 115,000 km on roads, and 35,000 km by air—that year I was away from Holland for 300 days.

The object of taking a TR2 on the Mille Miglia was to find the weak spots in the car when it was being driven flat out for more than 12 hours, and to break the car if possible. The machine, as prepared by Ken Richardson and a small team of mechanics at Standard's Banner Lane development workshops, was OVC 276, an early production model, which had left-hand drive. Although the engine was in virtually standard tune, the car was special in many ways, including the use of overdrive on all gears (with larger hydraulic pistons to operate that overdrive, and the overdrive switch on the gearlever knob.) In addition it had Koni rear shock absorbers, which meant that special chassis brackets were needed, and holes had to be cut into the floorpan to make space for these. Also specially strengthened 16 in. wire spoke wheels were fitted (the standard model used 15 in. wheels.)

Gatso was allowed to rev the engine to 5,500 rpm on this Mille Miglia car—which was

This was the very first 'works' TR2—as prepared to compete in the Mille Miglia of 1954. Ken Richardson is at the wheel of the left-hand-drive car, which has been built with two small aero screens.

well over the 5,000 rpm red line on the rev-counter—but with the larger wheels and Dunlop racing tyres the car could reach its top speed of around 120 mph with about 5,000 rpm on the clock.

The car was equipped with two aero screens, the bumpers were removed, and there was another very special (but not at all successful) fitting, as Gatso recalled:

> I had a special gadget fitted, so that we could take a pee while on the move—we had a tube through the floor—because I didn't want to leave the seat during the race. Fortunately we tried it out before the start of the race, when we noticed that it blew back, for there was pressure under the car, not suction! So we could only use it during the stops. All the officials kept saying: '*What* is Gatsonides doing?', and after we had gone, like a horse we had left our mark on the road. . . .

There was a typical Gatsonides run-up to the race, which involved the Tulip rally (in which he was to drive a Ford Anglia), the Rally Soleil Cannes, a recce for the Mille Miglia, and preparations for the Alpine rally which was to follow in July.

First of all Gatso drove a standard TR2 ('painted in a rather awful shade of pink') in the Rally Soleil-Cannes, which started at the Montlhéry race track near Paris and finished in Cannes, covering 750 miles and a number of special tests. There he was met by Ken Richardson in the Mille Miglia race car, both cars were then driven to Brescia in Italy where the race car was garaged, and the pink car was used to practise the route.

Richardson then took Gatso back to Milan airport, he flew to Holland, competed in the Tulip rally, won his class, flew back to Italy, and started the Mille Miglia the following morning, at 5.28 a.m.! Fourteen hours later the tired but elated crew were back in Brescia, having blasted their way around Italy in a completely open car, protected only by Cromwell motorcycle helmets, goggles, high-pitched horns and—in Gatso's case—a rubber apron to fend off any rain that might fall.

In the race itself Gatso and Richardson had little reason to speak regularly to each other—intercoms had not been invented, and there was a lot of wind noise which made conversation impossible.

> But it didn't matter, because finding the way was no problem, as there were so many people to mark the way. The only trouble with our green car was that if we were passed by a bigger red car—a Ferrari or a Maserati—all the people would walk out into the road to watch it, and ignore

Maurice Gatsonides (standing inside the door of the car) and Ken Richardson ready to start the Mille Miglia of 1954. In those days racing drivers didn't often wear overalls. Note that the experienced Gatsonides carried a spare pair of goggles around his neck. (Maurice Gatsonides)

the green car which was trying to keep up, behind it! Then, at railway level crossings, if a red car was coming the barriers would always be raised, but if it wasn't red the barriers could be lowered! In the end the 'works' TR2 had a copybook run, averaged 73 mph, and finished seventh in its class, 27th overall. Richardson, who had never suffered any doubts after the work he had put into prototype development in the previous two years, was delighted to report absolute-

Thumbs up from both occupants of the Gatsonides/Richardson TR2, captured in full flight in the 1954 Mille Miglia. The picture was actually taken by Klementaski, from the passenger's seat of Reg Parnell's Aston Martin race car. The traditional Mille Miglia numbering system meant that the TR2 had left the start, in Brescia, at 05.28. (Maurice Gatsonides)

ly no mechanical failings of any nature. A second car, privately entered by Coventry scrap dealer Leslie Brooke, and co-driven by Jack Fairman, was badly delayed when it ran out of fuel and the crew found that its spare fuel cans had been stolen—it eventually finished 64th.

As Ken Richardson later confirmed:

> Looking at these results . . . it was obvious that we had to have a works team, and so we entered the Alpine. I drove with Kit Heathcote who ended up working for me in the general running of the competition department as well as being my navigator and co-driver. His judgement and ability were second to none, and when he emigrated to Canada in 1957 I decided he was not replaceable and hung up my helmet for good.
>
> Lyndon Mills drove with Jimmy Ray, and Gatsonides with Robbie Slotemaker.

This was a real challenge for Standard-Triumph, and for Ken Richardson, as no-one except the drivers had any previous experience of long-distance rallies—and there were only eight weeks to get ready. Richardson had to listen to the recommendations of his drivers before finalizing the specifications. The mechanics (transferred from experimental work to prepare the cars) had never tackled this sort of thing before—and there was very little money to back an ambitious foray like this one. There were no service crews, though Frank Callaby's camera car carried a few spares, while the drivers carried a few 'get-you-home' parts in the boots of their cars. Some carried two spare wheels, but Gatsonides made do with only one.

Drivers' fees? None at all. All that a driver got was about £7 a day to cover his living expenses. His travel was paid for, and he was provided with money to buy petrol for the car on the event itself. Things have certainly changed a lot since then!

The Gatsonides/Richardson Mille Miglia car was refurbished for Richardson himself to drive, while two new cars—a left-hand-drive example for Gatsonides, and a right-hand-drive machine for Triumph's sales executive Lyndon Mills and British rally driver Jimmy Ray to share—were prepared alongside it. All the cars had wire wheels, ran with their bumpers removed and usually with their soft-tops retracted.

Three 'works' cars were entered because the factory had its eye on the team prize as well as individual honour, for in those days the team competition was a very prestigious award to be won.

Surprisingly, the winner of the 1954 RAC rally, Johnny Wallwork, was not offered the drive, and Jimmy Ray himself was a little surprised to be invited along. Jimmy, a confectioner from Lancashire, had started his rallying in cars like the Jowett Javelin and the Jaguar XK120, but had really made his name in British events like the London Rally, which he had won three times in succession driving Morgan Plus Fours.

Jimmy said:

> But until Denis Done, who worked for a Standard-Triumph agent in Chester, asked me along as third-man in a Vanguard in the Monte Carlo rally of 1953, I hadn't done any foreign rallies.
>
> It was after that, and after my British successes, that I met Lyndon Mills, and he asked me to go with him on the Alpine.

Gatsonides, naturally, carried out a very thorough recce of the route, using his pink car, but it was with hope rather than expectation that the team started the event from Marseilles. Like all French Alpines this was going to be a tough event, where the easy road sections were difficult, where the difficult road sections were almost but not *quite* impossible to drive on time, and where the speed tests sorted out a final result.

The Alpine—along with the Liège-Rome-Liège—were the two European events which the drivers enjoyed, but which they also feared. On the Alpine, to keep a penalty-free road record was everything—and exhausting—for to do this brought the reward of a *Coupe des Alpes* (Alpine Cup); to win one's class, or even the rally, on top of this was a bonus.

As in previous years, all of Europe's highest passes were used, some of them more than once. From Marseilles to Cannes, in four long stages, the cars had to tackle giant hills like

Practising for the Alpine rally in the mid-1950s involved pass-storming on gravel roads in the Italian Dolomites. The car nearest the camera was being driven by Maurice Gatsonides, who probably recce'd more assiduously than any other driver of the period. (Maurice Gatsonides)

the Stelvio, the Vivione, the Cayolle, the Izoard, as well as the exciting 'Circuit of the Dolomites'. With overnight halts scheduled for St Moritz and Cortina d'Ampezzo, with forays through Austria and into West Germany, it was going to be quite a trip.

It was a gruelling event with which to start a rallying campaign, and the following statistics tell their own story. 79 cars left the start but only 54 survived after the Circuit of the Dolomites. Only 37 cars finished, and of them only 11 collected Alpine Cups for clean runs.

The 'works' TR2s faced formidable opposition in their class, notably from Col. O'Hara-Moore and John Gott's Frazer-Nash Le Mans Replica, but their biggest challenge was inex-

Lyndon Mills (driving) and Jimmy Ray, hurling their TR2 up a pass in the 'Circuit of the Dolomites' section of the 1954 Alpine rally. Lyndon was a Standard-Triumph sales executive—Jimmy went on to win the RAC Rally of 1955 in a Standard 10. This 'works' car is carrying two spare wheels. (Jimmy Ray)

Jimmy Ray (driving this right-hand-drive car) and Lyndon Mills, in the Alpine rally of 1954. In those days 'works' rally cars usually ran in open-top condition. A hood erected was for cissies! (BMIHT/Rover Group)

perience itself. The drivers made no mistake in any of the speed tests—especially on the Munich autobahn sprint where many fancied machines failed—but Mary Walker (in her private TR2) crashed, the Mills/Ray car eventually retired with a failed wheel bearing, and Ken Richardson suffered a puncture and a broken rear spring, both of which cost time, but

Below *Almost the end of the road for coup-winning TR2 in the Alpine rally of 1954, with Gatsonides car about to tackle the final test*

Below right *Ken Richardson and Kit Heathcote on their first rally together—in TR2 OVC 276 in the 1954 Alpine rally. Because they rallied virtually without service support in those days, the TR2s ran with two spare wheels, one in the normal place in the boot, the other fixed to the boot lid. (BMIHT/Rover Group)*

he managed to keep going to the finish, albeit with 220 penalty points.

Maurice Gatsonides—who else?—confirmed his worth to the team by keeping his clean sheet, finishing sixth overall, and second in his class to O'Hara Moore's Frazer-Nash, while Triumph also took the much-coveted Team Award, on its very first event as a 'works' team.

But to win the team prize the 'works' cars needed the help of a privately-entered TR2, driven by Hans Tak (of Holland) and Joseph Kat, who was London-born but of Jewish descent. Although Tak was an experienced driver himself (he would win the Tulip rally in 1955, driving a Mercedes-Benz 300SL), it was Kat, the car's owner, who insisted on driving until he scared Tak so much that he threatened to leave the car and go home by bus!

From that moment, Tak was allowed to drive most of the difficult sections, and peace was restored, but Kat was driving on one descent from the Stelvio, when his car slid into the wall of a tunnel and bent the chassis rather badly.

'We were waiting at the Bolzano control, ' Gatsonides recalls, 'when this white Triumph arrived making a hell of a noise—his left wheelbase was 2 in. shorter than the right—it wasn't a square TR2 at all!' The local garage man suggested that it would take a day to repair, and since very little time was available it looked as if the team prize was lost.

Gatsonides takes up the story again:

> We drove on, thinking that the car was out of the rally. Suddenly I saw the white TR2 coming up fast in the mirror, I let it past, and we saw that all four wheels were running parallel again! Next time we stopped, I asked him how he had achieved this? 'We went to a blacksmith, and he said he'd try the brutal method—to give the car a horse pill! He fitted a steel cable just behind the front crossmember, the other end round a tree, then got us to reverse the car. The first two or three times, nothing happened. Then he shouted: "Faster, more power". We tried again, the cable broke—and the chassis was nearly straight again!' They managed to finish, not with a clean sheet, but at least the team was complete.

This had been a quite remarkable performance, and Ken Richardson had every right to be proud of his cars, his new and frankly makeshift team, and of his own efforts in putting it all together. Yet in 1954 there was more to confirm that the TRs were going to be formidable competitors in future years.

Even before the Alpine rally took place, Edgar Wadsworth and Bobby Dickson (the latter

Maurice Gatsonides (right) and Rob Slotemaker, along with the TR2 PDU 20 with which they won a Coupe des Alpes in the Alpine rally of 1954. (BMIHT/Rover Group)

This TR2, driven by Edgar Wadsworth and Bobby Dickson, was privately entered for the Le Mans 24 Hours race of 1954, but had some works support. It finished 15th overall.

a Standard-Triumph dealer from Carlisle) had entered a TR2 for the Le Mans 24 Hour race. Although the specification of this car benefited from the experience Triumph had gained a few weeks earlier in the Mille Miglia race, it was not a 'works', but a private entry. The Coventry registration of the car—OKV 777—merely confuses the issue for historians.

This, of course, was the titanic race in which Jaguar's new D-Types and Ferrari's latest 4.9-litre cars disputed the lead for the complete 24 hours, during which the TR2 kept plodding along. Although the weather turned nasty towards the end, the virtually standard sports car kept going, rose to 26th place after 12 hours, 17th after 18 hours, and in spite of a slipping clutch eventually came across the line in 15th place, having averaged 75 mph.

September emphasized all the successes of the early part of the year when Johnny

Johnny Wallwork, who had already won the RAC Rally of 1954 in his personal TR2, went on to win the London rally later in the year. His co-driver on that event was Willy Cave. (Johnny Wallwork)

Wallwork took his own RAC-rally winning TR2 to outright victory in the difficult London rally (Willy Cave was his navigator), and when no fewer that six TR2s started the long-distance Tourist Trophy race in Northern Ireland. Six cars started, six cars finished, and the cars walked away with the first *and* second team prizes too!

Writing in *Autosport,* W.A. McMaster commented:

> Brightest British performance, however, was undoubtedly that of the six Triumph TR2s which started. All six finished and between them won the SMM & T team trophy, and finished runners-up in the team competition.

John Bolster went further:

> Speaking of reliability, the performance of the TR2 Triumphs was excellent. That six of these moderately-priced cars started, and that every one of them finished at a respectable average speed, is beyond all praise; never has a team prize been so well deserved. It underlines the Triumph performance at Le Mans, and must have made many people reach for their cheque books.

One of the three-car teams was factory sponsored, this including Bobby Dickson/Ken Richardson in the Le Mans car OKV 777, McCaldin/Eyre-Maunsell and Lund/Blackburn; the best individual TR2 placing was by Brian McCaldin and Charles Eyre-Maunsell, who finished 19th overall. All in all, the TR2s finished 3rd, 4th, 5th, 6th and 7th in the Production Sports Car category.

It was a stunning end to a remarkable first season. It was no wonder, therefore, that Alick Dick (who had succeeded Sir John Black earlier in 1954 following a boardroom revolt) agreed to the setting up of an official Standard-Triumph competitions department. In December 1954 Ken Richardson became Competitions Manager, which should have been good news except that:

> I now had my own competitions department but, unfortunately, to staff it I had to take men made redundant from other parts of the company. They knew nothing about competitions and they were all in different unions, so when someone picked up the wrong tool there was an argument.
>
> I told the man he could forget unions while working in the competitions department. As a result, all Standard-Triumph's 17,500 workers were threatening to come out on strike. So I advised the directors to close down the competitions department at the works [Banner Lane] and to move it up to the service department at Allesley. This was far better, for I could have the six or seven men I wanted from the service department and there was no further trouble.

1955

Clearly this was going to be a good year for British rally teams. Not only was Standard-Triumph now to be set up with an official 'works' team, but BMC's team was also founded as well. At a stroke Britain became one of the most active nations in rallying—for the Ford and Rootes rally teams were already well established.

In its first full year, the Standard-Triumph 'works' competitions department was very active indeed. Teams were entered in the Monte Carlo, RAC, Tulip and Liege-Rome-Liege rallies (there was no French Alpine in 1955), along with a team in the Le Mans 24 Hour race, and another single entry in the Tourist Trophy.

But it wasn't all sweetness and light. To start the year Ken Richardson persuaded John Gott (the policeman who had co-driven Col. O'Hara More to so many successes) to join his team, but this arrangement lasted for only one event before Gott defected to BMC, where he became Marcus Chambers's team captain.

No other driver, it seems, was under contract, and the changes from event to event sometimes made little sense. Maurice Gatsonides, a previous Monte Carlo rally winner, was not

The Standard-Triumph 'works' team of Standard 10s before being flown off to the start of the 1955 Monte. Left to right: Johnny Wallwork, Jimmy Ray, Betty Haig, Mary Walker, Ken Richardson, Lyndon Mills, Kit Heathcote, John Gott and Ray Brookes. (BMIHT/Rover Group)

in the team for that event, nor for the RAC rally (he nevertheless ran Aston Martin DB2/4s on each occasion), but he did start the Tulip and Liege-Rome-Liege rallies. Johnny Wallwork, who competed in a 'works' car in the Monte Carlo rally of 1955 but on no other occasion, and Jimmy Ray, who only started three events during the year, both confirm that invitations came on an event-by-event basis.

No question about it, the new department was Standard-Triumph—*not* merely the Triumph—competitions department. Ken Richardson, who was a careful reader of the rules for motorsport (as Stirling Moss once quipped on another occasion: 'The competition begins when the regulations arrive . . .'), got approval for whichever model was most suited for each event, which explains why in 1955 Standard 10s were used on the Monte and the RAC rallies, but TR2s were the chosen weapons for the Tulip and the Liege-Rome-Liege events.

There is evidence, too, that publicity requirements sometimes coloured the entries—two different ladies' crews started the Monte Carlo and the Tulip rallies, while a journalist, Gregor Grant, borrowed a works car to compete in the Lyon-Charbonnières event.

For the Monte Carlo rally, where the event was to be settled by an acceleration and braking test before arrival in Monte Carlo, then by a strict regularity test circuit of 200 miles in the mountains, Standard-Triumph chose to enter no fewer than four of the new Standard 10s. These were compact and dumpy little saloon cars with 948 cc engines which, at first glance, appeared to be very unpromising material.

In hindsight, choosing to run modified Standard 10s in the *grand tourisme* category of the Monte looks strange, as all penalties incurred by these cars were increased by an eight per cent surcharge. John Gott, however, who was effectively the team leader, thought otherwise:

> I was therefore delighted when the offer of a wheel in the Standard team not only permitted me to compete in what may well prove the turning point on the road back to the great 'Montes' of pre-war, but also allowed me to prove most extensively a very intriguing little car . . . it was a wise decision to compete with modified Standard Tens.
>
> Basically, however, no modifications could endow the cars with the performance potential of Aston Martins, Salmsons and Alfa Romeos, with which, as far as the General Classification was concerned, we had to compete on level terms, or even with the more potent unmodified cars, such as Sunbeams, Jaguars and Daimlers, to which we had to give time.
>
> They succeeded, however, in transforming the little cars into real wolves in sheep's clothing, although no attempt was made to reduce weight by removing trimming, upholstery, etc. Indeed,

For the Monte Carlo rally of 1955, Standard-Triumph prepared a team of much-modified Standard 10s. John Gott, who later went on to be team captain of the BMC 'works' team, competed in this event for the factory team. His co-driver was Ray Brookes. (John Davenport)

The Gott-Brookes Standard 10 at a time control on the 1955 Monte Carlo rally. Note the drivers' names on the door of the car. (BMIHT/Rover Group)

the 'Monte' cars weighed considerably more, with their additional equipment, than the normal cars.

The engines were carefully assembled, balanced and polished and the compression raised to 8.4:1. Twin SU carburettors were fitted, and the camshaft profile slightly modified. All this raised the bhp to around 46, as compared with the normal 33. Good handling was improved by roll bars, and the brakes were carefully assembled and fitted with 'hard' linings.

The car behaved perfectly, the only attention required being a change of exhaust flange gasket, and had averaged just over 30 mpg, despite heavy use of the gears—maximum speed in top was 75/80 mph, and in third around 65 mph.

If all this seems to be a far cry from the 1990s, when rally cars have more than 300 bhp and four-wheel-drive, please remember that in the 1950s cars were much less sophisticated, and that organizers imposed complex handicaps to make every attempt to equalize chances.

Johnny Wallwork finally got his chance to drive in the team. Remarkably bright at age 78, he told me about his experiences. He remembers Ken Richardson as a forceful character, who always liked to get his own way:

> But I got on well with him. I always treated him with respect. He seemed to like that! He wasn't really a rally driver, but he went on to become one. . . . '
>
> When I got into the team, we were always very careful as to what we entered in which rallies . . . we'd read very carefully through the regulations to discover what the handicap system was, to find out where the key to the trophy cupboard was. . . .
>
> Now, you might think the Standard 8s and 10s were very slow, but they weren't—you'd be surprised. We could make them tramp on, especially with modified engines. I think it was possible to get 10,000 rpm out of them. . . .

[Surely that *has* to be an exaggeration?]

Jimmy Ray also got his second drive for the team for this event, once again as a second driver—this time to Johnny Wallwork. His first experience of the Standard 10 was in a pre-event trial:

> Johnny insisted on picking up the car early to try it out. He then started to do lots of little modifications. His first aim was to beat all the other cars—he wasn't a member of a team—team orders did not apply!
>
> He took the fan blades off, and after the event started I remember Richardson discovering this somewhere in France. He exploded: 'Who told you to take that off, those engines are carefully balanced. . . .' There was a big row, because they were two very strong personalities.

Johnny Wallwork's Standard 10 dips its nose under braking on one of the special tests of the 1955 Monte Carlo rally. These cars had modified engines with twin SU carburettors. Wallwork finished third in his class. (Johnny Wallwork)

Time to relax on the run down to Monte Carlo for the start of the 1955 rally. PRW 532 was John Gott's car, Ken Richardson is standing by the stones at the side of the road, and the Vanguard estate is Standard-Triumph's 'camera car'. (Mary Walker)

Although it would have been easier for the team to start from Glasgow on this, the first post-war 'works' entry in the Monte Carlo rally by Standard-Triumph, Richardson and Gott agreed that it made more sense to start from Monte Carlo. Not only would this allow the crews time to get used to the cars—and to 'shake out' any problems—on the way through France, but it would also allow them to look at the roads of the regularity section.

On this Monte the problem was not snow, but rain, floods—and the possibility of secret controls. In central France floods put paid to the chances of the Mary Walker/Betty Haig crew, though the other three cars made it to Vesoul, unpenalized. In the meantime, the irrepressible Wallwork had been having trouble with his car, as Jimmy Ray told me:

> I can remember that the passenger door kept opening as we were going along, so if you tried to sleep. . . . Johnny decided to fix that, and since he didn't speak French and I spoke a little, he asked me to find him an ironmonger's shop.
>
> When we found a shop, Johnny marched in, walked behind the counter, and sorted through

Final preparations for the 1955 Monte Carlo rally. John Gott crouches in front of Car '332', while Ken Richardson peers into the engine bay of the same car. The ladies' crew—Mary Walker and Betty Haig—get on with their own preparations to Car '295'. (Mary Walker)

Johnny Wallwork (in overalls) and Jimmy Ray, in the Standard 10 with which they took third in class in the Monte in 1955. The passenger's door on this car kept opening, at speed, so Wallwork bought a bolt from an ironmonger's shop to set things right! (Mary Walker)

all the boxes until he found the right sort of bolt, and some screws. . . . Then he bolted the door on the inside and solved the problem.

By the way, that wasn't all. I used to have a pair of Moorland slippers that came all the way up my ankles. I drove the whole Monte Carlo rally with those on. They were very comfortable to sleep in. We also bought a bottle of red wine, and put it by the heater so that it would get nice and warm. We had plenty of swigs on the way. . . .

In the end, though, the rally was won and lost on the regularity sections, where accuracy

Jimmy Ray (behind the wheel) and Johnny Wallwork prepare to leave the start of the 1955 Monte Carlo rally. John Gott, in bobble-hat, is talking to Ray through the window of the Standard 10. (Johnny Wallwork)

The Ladies' crew ready to start the Monte Carlo rally of 1955 in their Standard 10. Mary Walker is holding the passenger door of her car, while Betty Haig stands at the other side. This was the car which Jimmy Ray used to win the RAC rally two months later. (Mary Walker)

was more important than outright performance. Only two of the team cars crept into the 'top 100', though they went on to take third (Wallwork) and fourth (Richardson) in class.

This was not achieved, mind you, without Wallwork spending some time in jail in Monaco. Jimmy Ray told me that:

> When we got to Monte Carlo the final test was a race round the circuit. We were forbidden to practise, but after we'd all been to the Casino Johnny *wanted* to practise, so he got another car out of the garage. . . .

Hard a' port! Although this was Johnny Wallwork's 1955 Monte Carlo rally car, it was Jimmy Ray who actually drove it during the last test; Johnny had already been 'warned off' for illegal practising. (Johnny Wallwork)

Standard-Triumph's team of drivers, relaxing after the end of the 1955 Monte Carlo. Left to right: Kit Heathcote, Ray Brookes, Johnny Wallwork, Jimmy Ray, John Gott, Mary Walker, Ken Richardson and Betty Haig. (Johnny Wallwork)

I'd been in bed—I think I had a touch of 'flu—so I knew nothing.

When I got up the following morning I was told: 'You've got to drive the circuit test today. Johnny's in jail!' Apparently he'd being going round the circuit and the police picked him up and banged him into jail. They had to get John Gott to mediate, but he was only allowed out of jail on the understanding that he couldn't drive in the circuit race.

So I had to drive on the final test. Johnny Wallwork—the man you would most want to get

Three 'works' Standard 10s started the RAC rally of 1955. Not only did Jimmy Ray win the event outright, but the team won the Manufacturers' Team Prize too. Posing (left to right) are: Jimmy Ray, Brian Harrocks, Bobbie Dickson, Ian Roberston, Ken Richardson and Kit Heathcote. (Jimmy Ray)

you out of a crisis—but the man who'd probably got you into it in the first place! Life was never dull with him, I'll tell you.

This was a solid start, if not a sensational one, but no-one, not even the Tooth Fairy, could have forecast what would follow. Not only did Standard-Triumph enter three of the ex-Monte Standard 10s in Britain's own RAC rally, but Jimmy Ray won the rally outright, and yet another Manufacturers' Team Prize was added to the trophy cupboard.

In 1955 the RAC rally started from two seaside resorts—Hastings and Blackpool—and ended four days later in Hastings. It featured 11 special speed tests (some of them on racing circuits), difficult navigational sections in the West Country, Wales, Yorkshire and the Lake District and—as it transpired—a great deal of snow and ice. It was meat and drink for crews who were used to British navigational events, but the Continental crews were not at all impressed.

Standard-Triumph entered three Standard 10s—all in twin-carburettor 'GT' tune. Once again Ken Richardson led the team, and allocated another car to Standard-Triumph dealer Bobby Dickson, but on this occasion the third car (which had been driven by the ladies on the Monte) went to Jimmy Ray, and his regular co-driver Brian Harrocks. Mary Walker used her own TR2, in an attempt to repeat her success of 1954; amazingly, neither Wallwork nor Gott appeared in the entry list.

In the end, though, a car's performance mattered for nothing, for the weather was appalling, there were repeated traffic jams of rally cars attempting to reach controls, and it was ice-craft, car control and the experience of years of British rallies which brought the victory to the 33-year-old Jimmy Ray.

After heading its report: 'A Standard Triumph', *Autosport* summed up the event in this way:

> The fifth RAC British International Rally, which finished in Hastings last Saturday, will long be remembered as about the most difficult motoring event ever to be organized in this country. Wintry weather conditions played havoc with the 238 starters, and during the first series of special stages in Wales, crashes and lengthy hold-ups caused many to retire. . . .

Nor was it luck, or the good fortune of being at the head of the queue, which brought such a crushing victory to the 'works' team for the three cars started together—Numbers 213

Jimmy Ray needed all his skills to keep PRW 894 on the road during the 1955 RAC rally. Here the winning Standard 10 is sliding perilously on the snow of the Cadwell Park test. (Jimmy Ray)

Jimmy Ray (left) and Brian Harrocks, tired but happy, after winning the 1955 RAC Rally in a Standard 10. (Jimmy Ray)

(Ray), 214 (Dickson) and 215 (Richardson)—a long way down the lists. In many cases the drivers had to wind their way past other, stranded, crews, to reach their objective.

Jimmy Ray's co-driver was Brian Harrocks, a modest banker from Liverpool whose name was misspelt in the rally programme (as 'Horrocks') and who has never bothered to correct history in more than 35 years! Brian's first event with Jimmy was in the Scottish rally of 1953 in a Morgan, when the golfing-mad pair made sure that they were always first back to Gleneagles at the end of the day's run so that they could complete 18 holes before dinner!

'I think we were only given the detailed route half an hour before we set off, ' Brian Harrocks says, 'you'd see the hotels full of navigators and maps.'

Jimmy modestly recalls:

> The main reason we won is that we could get round those roads in slippery, icy, conditions—we were passing XKs and TRs like nobody's business. We had chains—and needed them. Brian helped me put them on, then he had to spend time in the boot, giving me extra traction over the driving wheels!
>
> It was my idea of what a rally really should be. I loved it, we'd done a lot of events like this. It was really like a long 'Yorkshire' to us. . . .

Even so, in those pre-computer days the results service was so slow that they could only guess at their performance, even to the very end:

> I think it was Raymond Baxter who kept telling us how well we were going, we kept listening to the BBC . . . I was amazed. I remembering having a dinner afterwards when we were told we had won. When we got across the finishing line we still had no idea. . . .'

At the end of an event which saw many crews sitting in their cars for hours waiting for icy lanes to be unblocked, Jimmy Ray's twin-SU equipped Standard 10 won easily, from a privately-entered TR2 (driven by Harold Rumsey of Pembridge), with Ken Richardson's car in third place. The gaps were wide, but as Bobby Dickson's car had kept going well, Standard also won the team prize. A young man called Keith Ballisat, who enters our story again in three year's time, won *his* class, driving a Morris Minor.

The *Daily Express's* noted motoring writer, Basil Cardew, summed up thus: 'Jimmy Ray, short and wiry, drove a car costing only £580 including tax. . . .' But Standard-Triumph nearly, so nearly, didn't record this victory, as Jimmy Ray told me:

> We went back to the factory on the Sunday to return the car. When we were just a few miles short of Coventry we had a fire in the car, with smoke coming out from under the dashboard. There was a bit of a panic when the car suddenly filled with smoke. We found out very quickly what had happened—we opened the boot where we'd been carrying a shovel to deal with the snow. The shovel had worn through against an electrical lead and shorted it. . . .
>
> Anyway, back at the factory we met the sales manager, Frank Hyam—and I remember him saying to us: 'Well, we have now sold our entire UK allocation of Standard 10s'

But the financial reward was tiny:

> 'Beforehand Ken Richardson had said that all money would be pooled—so I can't remember receiving a fat cheque, maybe £200 or so—but I do remember wanting the twin-carb kit for my own Standard 10. I mentioned this, and Standard-Triumph said: 'Well, you can have those for nothing'—and I think that was all.

By this time Standard-Triumph's competition success was really on a roll, and there was further excitement to come in the Tulip Rally. No fewer than five 'works' TR2s were entered—two of the 1954 Alpine cars, and three new cars—along with a Standard 10 for Maurice Gatsonides to drive. Because the Standard was entered in the 'Standard Touring' category, where modifications were not allowed, none of the victorious RAC rally team cars could be used, so NRW 953 was prepared instead.

Like many other rallies of the period, the Tulip was run on a complex handicapping system, which did nothing for the chances of the TR2s, but which favoured a crafty character like Gatsonides, in the slow but plucky Standard 10. Even so, there was no lack of incidents to keep the sports car crews awake.

The Standard-Triumph 'works' team ready to leave for the start of the Tulip Rally in 1955—one Standard 10 for Gatsonides, and four TR2s. Left to right: Maurice Gatsonides, Jimmy Ray, John Waddington, Kit Heathcote, Lola Grounds, Cherry Osborne, Bobbie Dickson and Ian Robertson. Gregor Grant's car—PKV 693—is not in this line-up. (BMIHT/Rover Group)

Gatsonides pedalled this unmodified Standard 10 into fourth place overall in the Tulip rally of 1955, though a favourable handicap helped him a little. Maurice, please note, is racing at Zandvoort wearing a shirt and tie—and no crash helmet! (Gordon Birtwistle)

Jimmy Ray and Cherry Osborne both set remarkably fast times on the Freiburg hill-climb test, but Cherry soon went off the road in the Vosges and lost a lot of time. Several of the cars were fast on the Col de la Faucille, but were still struggling against the handicap, but the real blow came near the Ballon d'Alcase, when Jimmy Ray's car suffered an accident:

I think we [Jimmy's co-driver was Johnny Waddington] were already in about third position, and we thought we could win. But we were going through a village in France, and there was this

Maurice Gatsonides (left) and T. St J. Foster won their class and took fourth place overall in the Tulip rally of 1955 in this Standard 10. Gatsonides, the deep-thinking Dutchman, had carefully read the regulations, and calculated that a standard car would be extremely competitive. He was right! (BMIHT/Rover Group)

Citroen 2CV, with a parish priest in it. We went one way, then he went the same—and we did this down the street until we hit each other, and that stove in my radiator and we were out of the rally. It was the most stupid thing to happen . . .

Later on Gregor Grant spun his car and bent a wing, the ladies' TR shed its bonnet and lost its screen, and all the cars were outpaced at Zandvoort by Rob Slotemaker's privately-entered TR2, and by John Gott's AC Ace.

In the end, however, Maurice Gatsonides won his class in the Standard 10, and after the handicap had been applied he finished fourth overall, just 21.2 seconds behind the winning Mercedes-Benz 300SL; the best of the TR2s—that of Ken Richardson—was second in class and 17th overall.

After this event Jimmy Ray accepted an offer to join the Rootes team, but gave me his lasting impression of the cars, and the characters, which he had met at Triumph:

I didn't leave because of any disagreement, it was just because of the support offered by Rootes; it certainly wasn't the money either. There were no fees!

I think Ken Richardson put a lot of effort into the team, those Standard 10s and the TR2s had everything possible done to them.

The following remark tells us a lot about Standard-Triumph's ideas on driver choice:

'But there was no such thing as a steady 'works' drive—he didn't take people on for, say, four or five events a year. Ken, well I don't think he was really a rally driver, and Bobby Dickson *certainly* wasn't a rally driver. . . .

It was this approach, no question, which caused John Gott to leave the team after such a short time; he settled happily in the BMC team for the next six years.

After the Tulip rally it was time for the team to prepare for a major new test—a team entry in the Le Mans 24 Hours race. In 1954 there had been interest in, and help for, Edgar

Triumph entered a team of three TR2s in the Le Mans 24 Hours race of 1955, and all had experimental disc-brake installations. PKV 374, carrying Number 68, was driven by Mortimer Morris-Goodall and Leslie Brooke. After spending 1½ hours in a sandbank, after Leslie Brooke had ditched it, it finished 19th.

Wadsworth's private entry, but in 1955 no fewer than three brand new TR2s were to be prepared, and run as a team from the factory.

The 1955 Le Mans race will always be remembered for the titanic struggle between Jaguar, Mercedes-Benz and Ferrari, and above all for the appalling crash, where Levegh's Mercedes-Benz 300SLR ploughed into the grandstands opposite the pits, and exploded into flames, killing not only the driver but more than 80 hapless spectators.

After the crash it was a joyless occasion, which detracted from the interest caused by Triumph's entry, and by the appearance of three prototype MGAs from the new BMC team. For Triumph it was not only an opportunity to reconfirm the speed and durability of the TR2s—in other words, to rub in the performances set up at Jabbeke, in the Mille Miglia, and in the 1954 Le Mans event, but it was also a chance for the factory to test different types of disc brake equipment. At this time disc brakes had not yet been adopted for *any* European series production car, though a lot of testing was going on. Both Dunlop and Girling were happy to see their equipment tested in public like this.

This was the line-up of TR2s for the 1955 race:

Car	Crew	Brakes
PKV 374	M. Morris-Goodall/L. Brooke (No. 68)	Girling front discs, 11 in. (Girling) Alfin rear drums. This car was originally a reserve entry.
PKV 375	K. Richardson/B. Hadley (No. 29)	Girling front discs, 11 in. rear drums, as above.
PKV 376	B. Dickson/N. Sanderson (No. 28)	Dunlop discs all round.

Harry Mundy, then the distinguished technical editor of *The Autocar*, and a long-time friend of Ken Richardson (for he had worked with him at ERA in the 1930s, and at BRM in the 1940s), summarized the cars' specifications better than I can. In the race preview he wrote that:

> Production cars are re-dressed with a tonneau cover over the passenger seat and a small plastic windscreen for the driver. Larger H6 carburettors have increased the engine power to 94 bhp.

This was the Richardson/Hadley TR2, which finished 15th overall at Le Mans in 1955. Note the use of front wheel disc brakes and oversize rear drums. (Gordon Birtwistle)

As part of the development programme the cars will be equipped with two versions of the disc brake. One car fitted with Dunlop disc brakes at front and rear will be driven by Dickson and Sanderson. The second car (drivers Richardson and Hadley) is equipped with Girling disc brakes at the front in conjunction with 11 x 2¼ in. shoe-type brakes at the rear; Alfin drums are used.'

In his after-race survey, Mundy then wrote that:

The Triumph TR2 had a rather interesting combination, which was used presumably to obtain first-hand experience for their possible use in future production. Two cars were equipped by Girling with 11 in. diameter discs at the front and shoe-type brakes also of 11 in. diameter and 2¼ in. wide, at the rear. The pair at the back had Alfin aluminium-bonded drums; neither pair had servo assistance.

An examination of the Girling arrangement on the Triumph reveals that the area of metal rubbed by the front discs is 109 sq. in. per brake, while that of the rear drum type is 80 sq. in. The third car was equipped with Dunlop disc brakes all round, and these had assistance in the form of a servo cylinder, operated through a connection from the induction manifold.

Apart from the braking installations, the use of 22 gall./100 litre fuel tanks, and the aerodynamic modifications made to reduce drag, these three right-hand-drive cars were virtually standard, though naturally they had been put together very carefully indeed at Allesley. The drivers all had a great deal of experience, though they were not necessarily in the top flight, or even the first flush of youth (Hadley, for instance, had raced the 750 cc Austin twin-cams in the 1930s).

Except for the occasion when Leslie Brooke ditched his car at the Arnage corner (which left it stuck in the sand bank for 1½ hours), the cars all circulated fast and reliably, and quite close together. By half distance they were running 18th, 19th and 27th, and by the end of the event they finished 14th, 15th and 19th.

As in the Mille Miglia (and, for that matter, as at Jabbeke) their top speed was about 120 mph, and the Dickson/Sanderson car, which finished 14th, completed 2,026 miles, and averaged 84.4 mph *including* all stops. The winning car (a Jaguar D-Type) had averaged 108 mph, and the fastest 2-litre car (a very special Bristol racing sports car) averaged 94.55 mph.

There was an interesting, and amusing, postscript. The third race car—PKV 374—had been finished off and shipped in a great hurry, one result being that its customs documentation was not completely in order. The result of this was that, whereas the two other cars were brought back to the UK without delay, the third car was not repatriated until May 1956, almost a year later!

In the aftermath of the Le Mans race disaster there was wholesale cancellation of events due to be held later in the year. This explains why Standard-Triumph had no chance to compete in an Alpine rally until 1956. Surprisingly enough, though, the Liege-Rome-Liege rally and the Tourist Trophy race *were* both held.

There were two 'works' TR2s in the Liege-Rome-Liege, which was its usual gruelling four-day marathon, with virtually no time for rest, and certainly with no overnight halt built in. Ken Richardson, who had never tackled the event before, decided to take 'Old Faithful' (OVC 276—originally built for the Mille Miglia of 1954), while Maurice Gatsonides (partnered by Georges Bourelly of France) got the second car. Gatsonides had competed in several earlier Liege events.

How to measure this phenomenal event? Probably by pointing out that the route covered 3,100 miles, through five countries, over 30 mountain passes, and occupied 90 hours. Only 56 of the original 141 crews reached the finish—and that only the first three finishers were unpenalized on the road sections.

It was an event which started and finished at Liege in Belgium, took in Germany, Austria, Italy and France, pivoted at Rome, but spent its most gruelling hours in the Italian Dolomites (twice) and in the French Alps. Although most roads were tarmac, there were gravel sections in the Dolomites, which makes Gendebien's winning performance in his

Mercedes-Benz 300SL even more remarkable, and special timed sections helped to sort out a result.

Through all this high-speed action, Richardson most remarkably lost a single minute on the road section (on the Vivione pass), while 'Gatso' lost four minutes at the same control. Although Richardson's TR2 broke its dynamo mounting on the last night, in the French Alps, after a quick repair no time was lost.

Back at the finish, in Belgium, it was party time *again* for Triumph, where TR2s had finished fifth (Richardson), and had taken the first three places in their class. As John Gott (who could not be expected to dish out unnecessary favours to Triumph) wrote in *Autosport*:

> The performance of Ken Richardson/Kit Heathcote in their TR2 was also outstanding as they were new to the event, although they had practised over a lot of the course before the event. Their fifth place represented the best-ever performance by a British crew, and they easily won the special prize for the best newcomers to the rally.

After that it was quite an anti-climax for Richardson and Bobbie Dickson to take one of the ex-Le Mans TR2s to compete in the Tourist Trophy race, and to finish 22nd overall.

1956

For the first event of 1956, the Monte Carlo rally, there was yet another change of emphasis in the 'works' team, as a mere two Standard 10s were joined by a massive team of six unmodified Standard Vanguards. There is no doubt that the marketing department had a hand in this, for the monocoque Series III Vanguard had only recently been launched, and on an event like the Monte, where strict time-keeping and experience counted more than car performance, it was thought that the cars should have a good chance. There was certainly no place for TRs in this event, as open cars were still banned from competing in the event.

Looking at the names of the crews confirms how much the marketing gurus had had their way, for there were *two* ladies' crews, while Peter Bolton had the BBC's Robin Richards as 'third man' to Arthur Slater in his car.

Except for another stirring performance by Maurice Gatsonides, this was a miserable outing for Standard-Triumph. Gatso, along with Marcel Becquart (who drove many of the difficult sections so that Gatso could look after the delicate timing which was needed), performed magnificently on the 150 mile Mountain Circuit, finishing eighth with all tyre treads worn away—and with ex-Triumph driver Jimmy Ray close behind, in a Sunbeam, in

The Vanguard III was an unpromising rally car, which means that Maurice Gatsonides' performance in the Monte Carlo of 1956—8th overall—was a magnificent achievement.

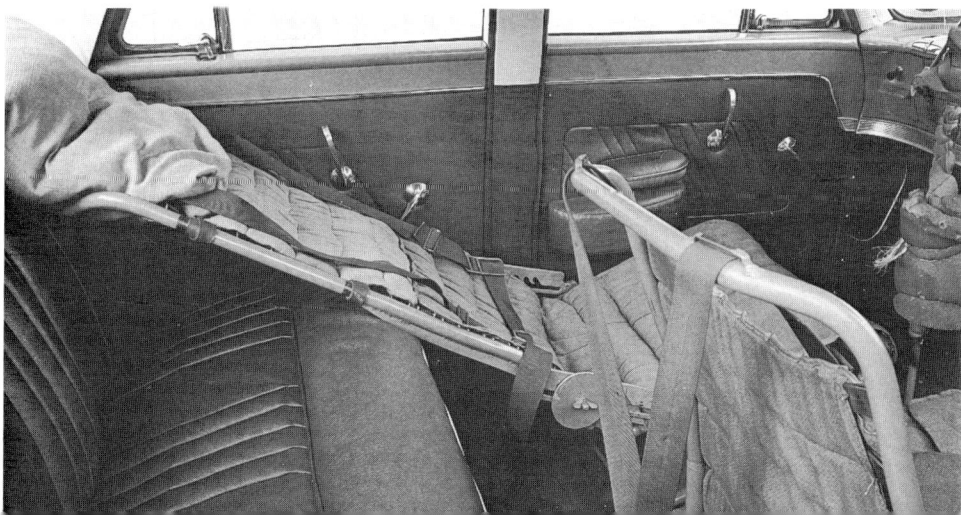

Maurice Gatsonides (left) and Marcel Becquart grappled with the newly-launched Standard Vanguard III to take eighth place overall in the Monte Carlo of 1956. Note the condition of the front tyres—both of them absolutely devoid of tread! (BMIHT/Rover Group)

tenth place. Gatsonides's Vanguard was fitted with lightweight Citroen 2CV-style seats for this event, and the car was so well-equipped that it won the 'Road Safety and Comfort' competition.

Gatsonides then took the self-same car (where it was re-registered) to compete in the Coronation Safari, later in the year, but in spite of having his own special form of disposable 'chains' for the tyres (short lengths of rope which the crew wound around the tyres to get them through patches of deep mud) he had no success.

In the RAC Rally, Standard-Triumph almost repeated its outright success of 1955, for a

How to save weight to go rallying! Gatsonides's Vanguard Series III for the 1956 Monte Carlo rally used modified Citroen 2CV seats. But were they comfortable? (BMIHT/Rover Group)

Paddy Hopkirk, in his first full 'works' drive for Standard-Triumph, urging his car around the pylons of an RAC rally test at Blackpool in 1956. (BMIHT/Rover Group)

team of Standard 10s not only won two classes (one for standard, and one for modified or 'Grand Touring' cars), but Johnny Wallwork still insists that he should have been the outright winner:

> Had it not been for bad management of the first test over in Lincolnshire, which was badly marked and bannered, I would have beaten the Aston Martin.

It was one of these events with two starting points, a seemingly interminable road section, a series of speed and manoeuvring tests all round the country, and difficult night-navigation sections in Devon, Yorkshire and the Lake District. For the intrepid crews of the Standard 10s, most of whom were seasoned campaigners in this type of event, it was familiar stuff, which explains the excellent results achieved. The team's new-boy, Paddy Hopkirk, started well, but sagged after the night-navigation sections—but this was just the start of an illustrious rallying career for the Irishman.

Paddy had arrived in the team after a great deal of lobbying from Standard-Triumph, Dublin, but his first 'works' appearance was delayed. Ken Richardson had offered him a ride in a Vanguard III as a last-minute substitute, but then found that the Monte organizers would not then accept a driver change at such a late stage.

The Autocar's Peter Garnier, ending his report, commented that:

> Finally, no paeans of praise would be complete without a bouquet for the robust and healthy little Standard 10s and Renaults and their relatively cramped crews. P. Cooper's 5th position in the general category with his Standard was a minor miracle.

Then came the Tulip rally, where the organizers imposed their usual handicapping system. This time, it seems, they slightly favoured the small-engined saloons, for in the end the top five cars all had tiny 800 cc engines. In an event where larger cars really stood no chance (compared with the Standards, Tak's Mercedes-Benz had a 'handicap' of 43 minutes on test times, so although he set a string of fastest times he could only finish 24th overall!), it was a fight to the finish between an Austin A30 (driven by the Brookes's, father and son), and no fewer than five 'works' Standard Eights, all of which had been specially built for this event.

Maurice Gatsonides was not contesting his 'home' event in a works Standard on this occasion (there was a near-clash of dates with the Safari, where he was currently practising hard), while Dennis O'Mara Taylor was a new name in the team.

Yet another fine performance by Standard 10s in the RAC rally—the 1956 event. The two 'works' cars were PRW 532 and PRW 894, the others being privately-prepared, but 'works'-supported. Outside the Imperial hotel in Blackpool are (left to right): Peter Cooper (1st in Class, 5th overall) and Geoff Holland, Tom Gold and June Gold, (second in class), Johnny Wallwork and Willy Cave (1st in class), Harold Rumsey and Peter Roberts 3rd in class. (BMIHT/Rover Group)

It was a long event, for there were eight starting points, the half-way rest halt was in Monte Carlo, the finish was back at Noordwijk in Holland, and along the way there were nine special tests—hill-climbs or circuit tests. Right from the start the 'tiddlers' dominated the results, and although the A30 and the Standards were among the slowest cars of all, they were outstanding in their category.

One of the Standards, driven by Cherry Osborne and Lola Grounds, expired on the Col de Grimone when the engine's sump plug dropped out, spilling all the oil on the ground. The other Standard-Triumph ladies' team (Jo Ashfield and Mary Handley-Page) then

The traditional 'team photograph' before a series of unmodified Standard 8s leave the Banner Lane factory to take part in the Tulip rally of 1956. The drivers are (left to right): Tom Gold and June Gold in their own privately-entered car, Paddy Hopkirk and J. Garvey, Dennis O'M. Taylor and Lew Tracey, Cherry Osborne and Lola Grounds, and Johnny Wallwork. (BMIHT/Rover Group)

Preparation of 'works' TR3s, at Allesley, for a 1956 rally. A check on photography dates reveals that it was staged after the victorious cars had returned from the Alpine rally. Car '412' was Gatsonides's car in that event, while SHP 520, in the middle of the front row, was Ken Richardson's car which had retired with a broken transmission. The car nearest the camera is SHP 990, which had competed in the Midnight Sun event. Note the bald rear tyre!

arrived at speed, slid on the resulting oil slick, and hit a tree, severely damaging their car's radiator. To keep at least one crew mobile, the radiator from Cherry's car was removed, fitted to the damaged car, and Jo Ashfield carried on, undaunted. Cherry Osborne, for her part, returned to Holland in a Press car!

Back at the finish, where the Standards beat the A30 on the final speed test at Zandvoort, the organizers were so shocked by the result that they ordered *all* these cars to have their engines stripped and measured. They even measured the length and gauge of the valve springs, to make sure that no cheating had taken place.

It was another great event for Standard-Triumph, where Ken Richardson's careful reading of the regulations had clearly paid off yet again. Standard 8s finished second, third, fourth and fifth overall, and once again a Manufacturer's Team Prize made its way back to Standard-Triumph's trophy cupboard.

By this time of the season one would have expected to see 'works' TRs in action, but not in 1956. There was no intention of repeating the 1955 visit to Le Mans, and handicaps had not favoured the use of TRs on other events. Almost immediately after the Tulip, however, a set of newly-prepared 'works' TR3s appeared for the very first time.

Although this was a European Rally Championship event, it was surely misguided to send cars to a specialized gravel event where the locals were almost bound to win? The results prove this, for Paddy Hopkirk's car could only finish fifth in the GT category behind four local drivers. A French girl, Annie Bousquet, made her only appearance in a factory Triumph, for unhappily she was killed in a racing accident (in another make of car) at Rheims a few weeks later.

The team's biggest effort of the year followed in July, when no fewer than four TR3s— three for the 'works' team, and one on loan to Tommy Wisdom—were prepared for the Alpine rally. On this occasion the cars were entered as 'Grand Touring Cars'—which is to say that they had hardtops fitted and external door handles, but little else to distinguish them from the usual TR sports cars.

The question has often been asked: 'How special were the "works" TRs?', to which Standard-Triumph always countered by stating that they were nearly standard cars, but beautifully prepared. This was laudable enough, but at a time when rival teams were introducing more and more modifications it struck several people as a rather complacent way of running a competitions programme.

Even so, it was a fact. In performance and roadholding they *were* almost standard, as Ken Richardson confirmed some years later. Writing in *Motor*, he pointed out that:

> The cars themselves were never very special. In fact we used to sell them off at the end of the season and I'd then ring up and ask for six more from the production line. . . . They were meticulously built in every way and then we would just leave them alone, apart from a bit of supertuning. Anybody could have a similar car to the ones we were using, which led to a terrible argument with a character who had asked us to prepare his car to works team standard.
>
> When he came to collect it he said: 'I bet it's not like your cars.' So I told him to choose any one of the works cars which were also new, and we'd take his instead, which more or less convinced him.

This ties up with what the various team drivers have told me, and it also explains the very anodyne article Ken Richardson wrote about rally car preparation for a motoring magazine in 1956; at first glance I thought he *must* have been hiding something. . . .

He also told an absolute untruth in this piece, which was that: 'We are in the fortunate position of having a consistent works team whose vast experience of Continental events . . .' —the fact was that there was really no such thing as a regular Standard-Triumph team at this stage, as drivers were invited on an event to event basis, and that they often had different co-drivers on each occasion.

On reflection, Ken probably had to stretch his script even to fill the page.

Even so, in the Alpine rally of 1956, which was as long, as fast, and as high as ever, Triumph achieved everything, and more, that they set out to do. At the end of the event three of the four Allesley-built cars had won *Coupes* for unpenalized runs (the only driver to

Below right *Paddy Hopkirk's TR3 climbing the Stelvio pass, on its way to winning a Coupe des Alpes in the 1956 Alpine rally.*

Below *The car is a 'works' TR3, the driver is Maurice Gatsonides, and the background is the unmistakeable back-drop of the Italian Dolomites. This was the Alpine rally of 1956. SRW 410 would go on to take fifth place in the Liege–Rome–Liege rally later in the year, then be sent off to race at Sebring in 1957. (Maurice Gatsonides)*

Above *A famous occasion—after the 1956 Alpine rally Triumph lined up no fewer than five TR3s which had all won Alpine Cups, and the team had also captured the Manufacturers' Team Prize. The three surviving 'works' TR3s, with their crews were (left to right): Tommy Wisdom and Ann Wisdom with SRW 992, Paddy Hopkirk and Willy Cave with SRW 991, and Maurice Gatsonides and Ed Pennybacker with SRW 410.*

Below *The cockpit of a 'works' TR3, circa 1956 or 1957. There are very few special fittings, the Hald Speedpilot is mounted on the transmission tunnel, the seats are absolutely standard, and there are no safety belts or other safety fittings. (Maurice Gatsonides)*

Below right *One of the trophies won by the 'works' team after the 1956 Alpine rally—the team award—was so big that the little girl could barely lift it! Ken Richardson (whose car retired) accepts the trophy, while behind him are the victorious drivers (left to right) Paddy Hopkirk, Joseph Kat (a private entry) and Maurice Gatsonides. (Maurice Gatsonides)*

miss out was Ken Richardson himself, whose car retired after a insufficiently tightened wheel flew off), and a total of five TR3s (of seven starters) had been 'clean' on the road.

Not only that, but the cars dominated their class, Gatsonides picked up his *Coupe d'Argent* ('Silver Cup') for three non-consecutive individual Coupes on the event, and the team prize was once again won.

In some way this was an easier Alpine that usual, for there were no fewer than four overnight halts in six sectors, but on the other hand the route struck deeply into dusty Yugoslavia—and only 34 of the 79 starters actually reached the finish.

In his annual review of the rally season, John Gott wrote in *Autosport* that:

> But undoubtedly *the* marque of the 1956 Alpine was Triumph, TR3s taking first place in their class (Gatsonides/Pennybacker), and the next four places, all with 'clean' cars, as well as every team prize open to them and the first Coupes des Alpes des Constructeurs ever to be awarded.

Amid all this excitement, the fact that a privately-owned Standard 10, driven by Dudley Barker, also won its class, was almost unnoticed.

Two months later there was a less ambitious entry in the Liege-Rome-Liege rally, which for the first time ever did not visit the Italian capital and which divided its difficult nights (four of them!) between Italy, Yugoslavia and France.

If you never did the Liege—I was involved on the team managing side on several occasions—you can have no idea of the sheer exhaustion it caused, for crews had to live in their cars for four days and nights, complete 3,200 miles and still go as fast as possible on some sections where a minute's lateness meant exclusion. Sleep had to be taken 'on the road', and food (if at all) snatched at the end of easy sections. It was equally exhausting for the cars, but the Triumphs were ideal for this sort of thing.

Two 'works' TR3s started—SRW 410 and SRW 991—one being 'official', the other on loan—and the result was that the French pairing of Liedgens/Rousselle finished fifth overall, and were second in their class to Claude Storez's twin-cam engined Porsche Carrera. It was a satisfactory end to a very busy and successful season.

1957

In the winter of 1956/1957 the world of motoring was stood on its head by the first Suez War, the consequent shortage of petrol, and the mass cancellation of motorsport events which followed it. For most British motorists, and especially for Standard-Triumph, 1957 would be a very restricted year. Because car sales fell back a little, so did the budgets available for motorsport programmes. The result was fewer entries in fewer events, so the cancellation of some important events must have come as a relief.

Petrol rationing was imposed in the UK in December 1956, and not removed again until May 1957, which made it inevitable that there would be no RAC rally in March, and that Ken Richardson's department would mark time during the winter months.

On the Continent, too, the fuel situation meant cancellation for the Monte Carlo rally, and the Alpine was cancelled for unrelated reasons. Just before the event got under way, the Italian authorities withdrew their permission for high-speed sections to be scheduled.

In the winter all this gave Standard-Triumph time to refurbish three of the 1956-model TR3s and convert them to 1957 specification before shipping them to Sebring, Florida, where Standard-Triumph USA entered them successfully in the 12-Hour sports car race.

In the whole of 1957, in fact, Standard-Triumph only entered a team of cars in three events—a trio of new green disc-braked TR3s were used in the Tulip and Liege-Rome-Liege rallies (plus a 'loan' car in the Liege), whereas three Standard 8s took part in the Swedish Midnight Sun rally.

The 'works' entry in the Tulip rally not only saw the disc-braked 1957 cars make their debut, but it also saw Johnny Waddington and Tom Gold—both of whom had already achieved great success in British rallying—in the team. Surprisingly, Maurice Gatsonides

Nancy Mitchell only drove for Triumph on one occasion—in the Mille Miglia of 1957. Here she poses with Ken Richardson, behind the TR3 which she drove in the event.

did not tackle the event in a Triumph, though he *would* be invited along later in the year to tackle the Liege-Rome-Liege.

Once again the regulations had been changed, and the number of starting points had been reduced to four, but there was unseasonably wintry weather (the TR3s had had their heaters removed to save weight, so the crews were not best pleased!) There were only five speed tests, four of them being hill-climbs, and the fifth a five-lap race around the Zandvoort circuit.

Unhappily, Tom Gold rolled his hard-top TR3 on the Freiburg-Shauinsland hill-climb

For 1957 the 'works' team invested in a set of new disc-braked TR3s. Their finest achievement was to win the Manufacturers' Team Prize in the Liege–Rome–Liege. Bernard Consten's third place-winning car is TRW 736, in the middle of this line-up. (Maurice Gatsonides)

test, which ruined any hope of a team prize, but the surviving team cars battled it out, including a nose-to-tail dice on the circuit in Holland, and were separated in their class by an AC Aceca at the finish. Waddington won the class, and also finished 12th overall on handicap.

Next, though, was an intriguing single entry in the Mille Miglia race, where Ken Richardson attracted Nancy Mitchell to race a TR3 in the production car category. In some ways this was a real coup, for 'Mitch' had just been crowned European Ladies' Champion of 1956—a season in which she had been driving for the BMC 'works' team!

The chosen TR3 was SHP 520—the white car which Ken Richardson had used in the Alpine rally of 1956—but for this occasion it was equipped with a monstrous 25 gallon fuel tank, and had the latest Girling front-wheel disc brakes. It would be Nancy's second Mille

Above and below *Before . . . and after, the Liege–Rome–Liege of 1957 are Maurice Gatsonides (left in the 'before' shot) and Andre Jetten. They finished fifth, with a dirty but otherwise unmarked car. (Maurice Gatsonides)*

Miglia, for in 1956 she had shared an MG MGA with Pat Faichney, and had finished 72nd overall.

In 1957 Nancy and Pat started well, and were averaging 86 mph before they arrived at Pescara, having completed 400 miles. Then—disaster—the car spun at a level crossing, hit a pile of straw bales, and some wire from one of those bales punctured the radiator, letting out all the water. As Nancy wrote, many years later, in the TR Register's newsletter:

> The year before we had had an open MGA with aero screens. The axle ratio [of the TR3] was raised, wouldn't pull the skin off a rice pudding, but we needed speed, and boy we had it.
>
> Anyway we arrived at the level crossing at Pescara—it came up miles and miles before my speedo trip [the speedometer gearing had not been altered at the same time as the axle] . . . All the water poured out of my radiator. Luckily there was a garage nearby. Took hours and hours to repair the leak. . . .

As in 1956 the team entry in the Midnight Sun rally was a complete waste of time. Could *anyone* at Standard-Triumph seriously have expected British drivers in under-powered Standard 8s to match the skills of Scandinavian drivers in front-wheel-drive DKWs and Saabs? The fact is that Standard's money was wasted, for there were no results to boast about.

Before the end of the year, therefore, the team had to get its reputation back—and it did this in no uncertain way by sending four TR3s on the Liege-Rome-Liege marathon, and putting up the best performance ever achieved by this car.

Maurice Garot, the arch-organizer of the Liege, had excelled himself on this occasion, setting crews to complete more than 3,100 miles in less than 90 hours, turning the event back at Zagreb in Yugoslavia, and making sure that crew—and car—fatigue was more important than accurate time-keeping and regulation-watching.

On the Liege, in fact, entrants were encouraged to produce very special cars, and many did so—except, that is, Ken Richardson, who did not believe in that sort of thing. TRW 735, TRW 736 and TRW 737 were the usual type of TR3 'endurance cars'—though on this event his secret weapon was to employ European crews throughout.

It was a race really, and a battle to stay awake, rather than a rally, so experience worked wonders. Maurice Gatsonides had practised longer and more carefully than his team-mates, but still had to allow his co-driver to drive some sections. On one of these he must have wished he was behind the wheel when the co-driver slid off in the fog at the top of the Col du Galibier, but there was no damage.

Half way through the event, as 62 cars survived the Yugoslavian and Italian sections, the Storez/Buchet Porsche Carrera was already well in the lead, but the four 'works' TR3s were in second, fourth, sixth and 12th places. A day later one of those cars had gone out—the old Mille Miglia car, driven by Leidgens/Dubois had lost its fourth place, and succumbed to electrical and engine bearing failures—and by the end of the night in the French Alps the Schlesser's Mercedes-Benz 300SL rocketed up into second place.

Nothing, however, could diminish the great result set by the three identical-looking green Triumphs—third, fifth and ninth overall, and yet again the winners of a much-envied Manufacturer's Team Prize. Except for that elusive outright victory—which a British car had not yet achieved on the Liege-Rome-Liege—no-one could have asked for more.

Standard-Triumph, and Ken Richardson, however, *were* looking for more in the future. In the next few years there would be a new version of the TR3,   a new small saloon from Triumph—and a new engine with which to attack the Le Mans race. By the end of 1959, indeed, 'works' Triumphs looked different in almost every way.

Chapter 3

TR3As, twin-cams and Heralds

1958-1961

In the next few years, Triumph's motorsport effort seemed to be in a period of almost continuous change. TR3s gave way to TR3As, Heralds took over from Standard 10s and Pennants, and Triumph also returned to Le Mans, not only with a new type of engine, but with a new shape of car.

That was the good news. The bad news was that Standard-Triumph's financial fortunes collapsed in 1960/1961, the company was eventually taken over by Leyland Motors, and the competitions department was closed down in 1961. Ken Richardson, who had run the department since it was set up in 1954, left the company almost overnight, bitter at his treatment.

1958

The 'works' team started the season by taking delivery of two sets of new cars—four new green TR3As and four new Pennant saloons. As ever, Ken Richardson wanted to be able to read rally regulations, hedge his bets, and to enter cars best suited for every occasion. The lasting mystery is why the VRW 222 registration number was never applied to a 'works' TR3A in this set.

Below Annie Soisbault was already the French Ladies' Rally Champion before she was signed by Standard-Triumph. In 1957, using this privately-owned TR3, she had won the Coupe des Dames in the 1957 Tour de Corse. (BMIHT/Rover Group)

Below right Ken Richardson celebrates the signing of the French champion, Annie Soisbault, for the 1958 rally season. The TR3A, naturally, is parked within sight of the famous Paris landmark, the Arc de Triomphe. (BMIHT/Rover Group)

Maurice Gatsonides and Marcel Becquart made another stirring drive in the Monte Carlo rally of 1958, using this new TR3A.

For the Monte Carlo rally, where the two-seater hardtop TR was at last eligible to take part, Ken Richardson entered all the new TRs for the 'A-team', along with a Standard 10 for Johnny Wallwork to drive. The Standard was in what we now knew as the usual highly tuned form, complete with twin SU carburettors, overdrive and larger drum brakes. Three TR3As started from Paris, though Gatsonides chose to start from The Hague (in Holland), and Johnny Wallwork started from Glasgow.

This was the first time that a tall, elegant, but rather enigmatic French girl called Annie Soisbault drove a 'works' TR—she had already made something of a name for herself, using her own TR3, in French events, and had been signed up for a complete season. Annie was allocated a left-hand-drive car (VRW 219), though Gatso had to make do with a right-hand-drive car as usual.

In the same event, incidentally, there were three Standard Ensigns, which had been prepared by the factory (VHP 721, VHP 722 and VHP 723) but which were entered by the British Army Motoring Association.

The Monte Carlo rally of 1958 will always be remembered for the simply horrendous weather in which it took place. There was a complete blizzard, or white-out, in central and southern France—so of the 303 cars which started, only 59 cars reached Monte Carlo after the concentration runs, and only 38 completed the Mountain Circuit. Entries from Paris (85 starters), Munich and The Hague were practically wiped out by snow and ice in the Saint Claude area of France, though many of the Glasgow starters struggled through.

Three of the TR3As retired, either off the road, or simply out of time, but two team drivers—Maurice Gatsonides and Johnny Wallwork—struggled through. On arrival Wallwork had 210 penalties, while Gatsonides (Gatsonides of all people!) had amassed 980 penalties, and was way down in 58th place of 59 survivors:

Gatsonides recalled:

> I was driving in the Vosges mountains. There were steep hills, and there was thick fog. I went round a fast corner and there was a DKW standing across the road. It happened to be Rob Slotemaker, he had blocked the road in the snow—I hit his car. I couldn't miss him!
>
> The front end of the TR3A was damaged. We had a repair done by a blacksmith in the mid-

dle of the night at Saint Claude, because the chassis was damaged. We managed to get that battered car to the control but we were nearly at our maximum lateness.

Then, in the Mountain Circuit, Gatso and Marcel Becquart made the drive of their lives:

> Before the start we had 30 minutes to fix the car. In that time we managed to fit a tractor head-
> lamp which we had bought beforehand. With the help of the police who allowed us to drive the
> wrong way up a one-way street, to get to a little garage to get the work done! This headlamp was
> fitted, with a drill through the bonnet.
>
> Becquart drove the car in the Mountain Circuit, and I was timekeeping. I pushed him hard
> because I knew every inch of the way, and in any case he is a Savoyard. He kept shouting back
> at me: 'Don't push me any harder, I can't go any faster. . . .'

In what was, to all intents and purposes, a 655 mile night-and-day race over ice-covered
roads, this amazing French-Dutch combination pulled right up—from 58th to sixth overall.
Although well back in the starting order, the car was eighth to return to Monte Carlo, and
had set the third best performance, overall, on the Mountain Circuit, behind the two cars
which took first and second place in the rally.

Wallwork's drive, with Tony Beaumont in the Standard 10, was almost as amazing.
Johnny told me:

> Very few people got through to Monte Carlo with a clean sheet. We Glasgow starters were lucky
> because before reaching Chambery we went across the Central Massif—Clermont Ferrand and
> Le Puy—but those people who started from other points went to the Haute Savoie first and got
> horribly stuck.
>
> I was competing in the same class as the Dauphine which won, and it was the first time
> Michelin had produced studded tyres for them. The driver lived locally, so he had two advan-
> tages—the second was that his pals were out looking for the secret regularity checks, to tell him
> in advance. . . .

Even so, Johnny was another 'works' driver who put up a splendid performance in the
mountains, and clawed his way up from 24th to 13th place. Cyril Corbishley, who drove his
own Standard 10, but in the standard category, took second place in his class, and 15th

*Johnny Wallwork and Tony Beaumont not only battled through the blizzards to reach Monte Carlo
in 1958, using this Standard 10, but finished second in their class. This shot was taken on the Col
du Turini—and by the looks of it all Beaumont is in the back seat to ensure maximum traction on
the icy roads. (Jimmy Ray)*

Pre-RAC rally preparation, at Allesley, of the Standard Pennants which so nearly won the 1958 event outright. Of four cars entered two of them—driven by Ron Goldbourn and Tom Gold—finished second and third overall. (BMIHT/Rover Group)

place overall.

After such a traumatic event as this Monte, Standard-Triumph might have expected better weather for the RAC rally of Great Britain, but in March 1958 conditions were even worse than they had been in 1955, when Jimmy Ray had won the event. Once again there were mountains of snow, acres of ice, and the need to visit controls which invariably seemed to be placed at the top of steep hills.

As in previous years, Ken Richardson chose to use nimble small saloons rather than the TR3As, this time building up four new Pennants. These were revised versions of the Standard 10s, with new front and rear styling, but the running gear was essentially unchanged, which meant that all his department's hard-won experience could be used one again.

No fewer than four Pennants started the rally (in one of which was a bespectacled young co-driver called Stuart Turner, who would go on to be a distinguished team manager at BMC, and later at Ford) along with the ex-Monte Carlo rally Standard 10. These were entered in the Grand Touring Category, which meant that they were pitted against all the Jaguars, TRs, and Morgans. Once again there seemed to be little logic in choosing drivers, for Gatsonides and Wallwork, who had driven so well on the Monte, were dropped, whereas Cyril Corbishley was given a works car on this occasion! Wallwork reacted by entering his own Alfa Romeo.

By tradition, almost, the RAC rally was a real challenge, for a complex route, which included difficult navigation sections and 20 speed tests, was made worse by very wintry weather. As in 1955 and 1956, it was not the fastest drivers who won, but those with the most experience of bad weather, with encyclopaedic knowledge of Britain's minor roads, and sheer gritty determination to keep going at all costs.

During the snowy Welsh sections both Ron Goldbourn and Tom Gold (in 'works' Pennants) kept a clean sheet (the only cars to do so), which was miraculous, so they led the rally outright at the half-way halt in Blackpool. Conditions were even worse in the Lake District and the Borders, but all the team cars kept going somehow. Paddy Hopkirk's Pennant suffered axle problems near Kelso, but the Irishman diverted to Tweedsmuir Motors (a local Standard-Triumph dealer), found enthusiastic staff on hand, and persuaded them to donate a secondhand axle from a Standard 10 which they had in their showrooms.

The two Pennants and their crews, who took second and third places overall in the RAC rally of 1958. Nearest the camera are Stuart Turner and Ron Goldbourn, with Tom Gold and Willy Cave by the car behind them. (BMIHT/Rover Group)

That wasn't the end of his troubles, for his car's windscreen shattered on its way from Brands Hatch to the final control at Hastings.

Ice-master Peter Harper (Sunbeam Rapier) went even better than the leading Pennants in the Lake District, but Messrs Goldbourn and Gold finished second and third overall. By finishing third in their class Cyril Corbishley and Phil Simister (Standard 10) also helped the company to take yet another team prize.

Hastings in March, on a windswept promenade, where Tom Gold's Standard Pennant is on its way to third overall.

In 1958 the RAC Rally was a very good year for Standard-Triumph. Not only did the Pennants almost sweep the board, but this car, driven by Cyril Corbishley and Philip Simister, finished sixth overall. (Cyril Corbishley)

The season then hotted up considerably, for a few days later Annie Soisbault had her first success in a factory-built TR3A, by finishing sixth in the Lyons-Charbonnières rally, while three other 'works' TR3As were sent to Ireland where Paddy Hopkirk won the Circuit, Desmond Titterington finished second and—guess what—the cars also won the Team Prize. It was getting to be a habit.

The budgets must have stretched a long way in those days, for Triumph then entered no fewer than six cars in the Tulip rally—one of them being a Standard 10 for Gatsonides to pitch against the handicapping system. This was the event in which a tall and slim young man from Shell Motorsport—Keith Ballisat—had his first drive in a 'works' TR3A. The oddity here was that Triumph's fuel contract was with BP, though as Keith told me: 'It was

Autosport's Editor, Gregor Grant, co-driven by photographer George Phillips, sets out on the Lyon–Charbonnières rally of 1958, in the left-hand-drive TR3A more usually driven by Annie Soisbault.

VRW 221 had already finished second overall on the 1958 Circuit of Ireland before Ron Goldbourn and Stuart Turner used it to win their class, and finish tenth overall in the Tulip rally. (BMIHT/Rover Group)

a little bit awkward, but Shell shrugged it off.'

Keith's memory of his first year with Triumph is that there was very little factory support on the events, and certainly no service car umbrella:

> On that Tulip rally of 1958 I don't remember a service car at all. In the Alpine which followed there *was* some support, but we were met only once every 12 hours or so.

Keith also remembers Ken Richardson as a hard-working team manager who had his own fixed ideas of the way things should be done:

> We had very little influence on the specification of the cars, we normally only picked them up

The final circuit speed test of the 1958 Tulip rally, with Annie Soisbault's 'works' TR3A leading a private TR3A and one of the other 'works' TRs. Annie finished third in her class on this event. (BMIHT/Rover Group)

Maurice Gatsonides drove this ex-Tulip rally Standard 10 to a creditable finish on the Austrian Alpine rally of June 1958. (Maurice Gatsonides)

the day before the event. But at least we always wrote a critique afterwards, to list what had gone wrong, and what could be improved. We *certainly* didn't have individual cars tailored to our own requirements.

There were no seat belts, no roll cages, no nothing—we never thought about it. We only used overalls to go racing—but for rallies we just wore jumpers and slacks. There was no question of crash helmets on special stages. We didn't have special seats either—these were virtually standard motor cars, carefully screwed together. We had stiffer dampers, but nothing was done to the engines.

Cyril Corbishley pedalled the Standard 10 as hard as he could all round the Alpine Rally of 1958, including this gravel section over the Passo di Vivione. It was an untiring effort, though Cyril was eventually beaten by the unrelenting time schedules. (Cyril Corbishley)

There were no fees—only expenses—and no contracts, only verbal agreements. We only got £7 a day towards living expenses, which seemed to be enough in those days. There certainly wasn't a fixed programme—I don't think Standard-Triumph *had* a programme in those days, which meant we didn't get much notice of when we'd be needed. That explains why some of us had to miss some of the events—in 1958, for example, I simply couldn't do the Liege, because I couldn't get time off. It was the same for some of the other drivers.

By half distance Johnny Wallwork was up to tenth place (on handicap, remember) but he had to retire near the end of the event when two wheels collapsed; no-one was admitting to anything, but I am sure this happened after Wallwork had left the road and hit something! The rest of the team had all manner of problems, with Annie Soisbault's co-driver making navigational errors, Keith Ballisat losing nine minutes on the road, and with Ron Goldbourn's TR3A hitting a Porsche on a special stage, vaulting into a field, but carrying on to win its class.

For Standard-Triumph, however, the major event of the year was always the Alpine rally, and for this occasion Ken Richardson produced a modified engine—this being the very first of the 2.2-litre units. It did little for the top-end power, but it gave significantly more torque. Richardson hoped that it would be enough to make his cars a match for the Austin-Healey 100-Six team—and he was right.

As ever, the Alpine was a high-speed road race which started and finished in Marseilles, took in night halts at Brescia and Megeve, and included virtually every high pass in the Alps and Dolomites, all of which were tackled at near-impossible target average speeds. In his *Autosport* preview, John Gott described the route as having: 'probably the toughest last "Alpine" day ever.'

In the event, it was yet another outstanding performance by the Triumphs, where Keith Ballisat took fourth overall, and won his class, with Desmond Titterington finishing third in class. Not only did Keith win a *Coupe* (one of only seven drivers to do so) but he was accompanied by a French journalist, Alain Bertaut, whom he had never met before the event!

It was on this event, however, that Maurice Gatsonides' disillusionment with the team began:

Desmond Titterington, new to the 'works' team in the Circuit of Ireland in April 1958 (where he finished second overall), was retained for the Alpine rally in July, and hung on well to finish third in his class, and eighth overall. VRW 221 had already won its class in the Tulip rally.

At a time check near Barcellonette we had to repair the brakes. I knew there was a leak some-where. We found fluid coming from the left rear wheel, and found a leak where the pipe joined the cylinder on the rear brake.

What had happened was that the shock absorber at that side was rubbing on the brake pipe, and had caused the leak. That car had been crashed on the Tulip rally a few weeks earlier, and had not been re-prepared properly, because the pipe should not have been touching the shock absorber.

There was no service crew available, and we only had 15 minutes to spare, but I was driving for a Gold *Coupe*. I decided to cut off the rear brakes completely—by putting a piece of inner tube in the T-piece—and drove on only with front brakes.

On the way to the Col d'Allos we drove as quickly as we could—every time I needed the brakes I shouted 'Brake' in Dutch, when I hit the foot pedal and Andre Jetten, my co-driver, heaved on the hand brake.

Then we found this corner with gravel on it, we were too fast, we had to choose between a big drop into a river, or a hard rock on the other side . . . I chose the hard rock, and the car finished up half a metre shorter on one side. I was *very* unhappy about the preparation, for it probably cost me my Gold *Coupe*. But it could also have cost me my life . . . that was just one of several things which caused me to break with Ken Richardson. . . .

This was the event in which Paddy Hopkirk's TR3A punctured on the climb of the Stelvio, after which Paddy kept on driving to the top, cooked his engine, and was never asked to drive for Ken Richardson's team again. Though Ken Richardson would never admit this, it was a tremendous loss for Triumph, for Norman Garrad soon invited Paddy to drive Rapiers for Rootes, and from 1962 Paddy then joined the all-conquering BMC team.

It was also an event in which Triumph discovered that Annie Soisbault, who had still not teamed up with a regular co-driver, was—how shall I put it—rather wayward. As a dri-ver she was very fast and capable, but as a person she seemed to be somewhat disorganized (Valerie Domleo, who co-drove for her on the 1959 RAC rally, described Annie as 'very vague'). Maurice Gatsonides remembers her as 'not in the same class as Pat Moss, but she had a good reputation in France, where she was already the tennis champion.' Perhaps John Gott, writing after the event in *Autosport,* put it best of all:

> Annie Soisbault had a most indifferent rally, being late at the start, early in the Allos regularity test, and finally retiring after an accident on the Stelvio. . . .

The last incident, by the way, was caught on film for all time by Shell, which was producing a magnificent visual record of the event.

John Gott (who was also team captain of BMC, whose Austin-Healeys had been beaten by the TR3As) went overboard in his reaction to the latest car, and also leaked the exis-tence of a new engine:

> Possibly the most intriguing cars were Ken Richardson's Triumph TR3As, with the new 2.2-litre engine, an interim measure until the introduction of the even newer twin-cam. The extra cc obtained by the use of a larger liner, produced not only more bhp, but also better torque, result-ing in improvements of over 10 seconds per lap round Monza and over 20 seconds up the Stelvio, where Ballisat's car made B.T.D., as it did also on Mont Ventoux. . . .

For a normally logical thinker like Gott, these were sweeping statements, as the new engine was not the only improvement to have been made since 1956. There was also the fact that Dunlop had finally produced radial ply tyres for their teams to use, the TRs had disc front brakes (they had only used drum brakes in 1956)—and the fact is that Keith Ballisat was an outstanding, and very brave, new driver in the team.

Even so, the cars were definitely faster and more flexible than before, as solid perfor-mances in the Liege-Rome-Liege which followed a few weeks later also proved. On this fast and rocky marathon, which was as difficult as ever, and which spent 932 miles in Yugoslavia, where most of the mechanical carnage took place, only 22 of the original 98

Keith Ballisat in the Italian Dolomites on his way to winning his class and finishing fourth overall in the Alpine rally, using a 2.2-litre engined TR3A. The engine was newly-developed for this event, and the cars defeated the Austin-Healey 100 Six 'works' cars. This car used disc wheels, but . . .

starters reached the finish. Two of the four 2.2-litre 'works' TR3As made it, and a privately-entered TR3A (one of three BAMA cars which had been prepared at Allesley) went on to win the 2-litre class.

Two works TR3As finished in the top ten, but on this occasion Gatsonides (repeating his fifth place of 1957) was beaten by Pat Moss's Austin-Healey 100-Six, which was an ominous taste of what would happen more and more often in 1959 and 1960.

It was the last time the 1958 set of TR3As was ever used in a rally, for new cars would be prepared for 1959. Annie Soisbault's fight for the Ladies' series had not gone at all as well as

. . . Maurice Gatsonides had centre-lock wire wheels on board for the Liege–Rome–Liege, where he finished 5th overall, the best of the 'works' Triumphs. (Maurice Gatsonides)

According to me, this was definitely a 'works' entry in a major rally—the proving run of the proto-type Heralds took in Cape Town to London, via the Sahara desert, in the autumn of 1958. Timbouctou? Now isn't that almost a familiar sign?

Standard-Triumph had hoped—there had been a lot of publicity put behind her appearances—but this had been a learning year (she had done very little Championship rallying before being signed up by Ken Richardson at the end of 1957), and more was expected of her in 1959.

There is just space to mention one long-distance expedition which wasn't a 'works' motorsport effort, but certainly qualifies as one! John Lloyd takes up the story:

> One day when I was with prototype [Herald] cars, testing at MIRA, Harry Webster and Martin Tustin arrived, took me on one side, and said: 'How would you like to take a couple of these cars out to Africa and back?'

That was the start of a scheme to ship cars to Cape Town, then drive them all the way back to London, by way of Equatorial Africa and the Sahara desert, as a proving exercise. Two Herald prototypes—a saloon and a coupe—a Pennant Companion (estate), and an Atlas minibus made the trip, with John Lloyd, a cameraman, six fitters and (from Kano in Nigeria) Richard Bensted-Smith of *The Motor*.

The convoy left Cape Town on 18 October 1958 and, not without dramas, arrived on the Mediterranean shore, at Tangier, in mid-December, 9,650 miles later. After a ferry crossing to Gibraltar the crew couldn't wait to get home, and drove non-stop, for 46 hours, to get back to England.

Although all cars made it, there were many breakages to be mended which, in a way, taught lessons which helped make the production cars stronger and more reliable. The Pennant was rolled, the Atlas spent most of the time with its engine almost boiling, while the Herald broke half shafts, rear suspension radius arms, wishbone lower mounts, and the front suspension towers slowly bent towards the centre of the car.

Even so, by the end of the trek, most of the team wondered if this had been the best way to test the incredibly overloaded cars. As John Lloyd summed up:

'We sometimes wondered what we were trying to prove.'

1959

Although Ken Richardson planned a full rally programme for 1959, he was about to be swept up in Alick Dick's growing obsession with Le Mans. Competing in the Alpine or the

Liege-Rome-Liege might have been important to Ken, but as far as Alick Dick and Harry Webster were concerned, the most important motorsport project of the year was the return to Le Mans, with a new twin-cam engined car.

This was also the year in which the TR3A began to lose ground to its rivals. The introduction of the 2.2-litre engine in 1958 was the last significant improvement ever made to the team cars, which meant that they were increasingly confined to class wins, or to gallant performances where brave driving meant more than car performance.

Richardson himself once said that the : 'Competition cars were virtually standard . . . I can't understand the speculation about this. We had absolutely no reason to race-tune the TRs. They could beat the opposition without it, and did for years. . . .' Those years were 1954 to 1958, for after that it was always a struggle.

This complacency, and the stagnation of the TR3A was inexplicable, for the latest homologation rules would have allowed Weber carburettors to be used, aluminium body panels to be fitted, and all manner of other performance-raising extras (like four-wheel disc brakes) to be added to the specification. Yet none of this was done.

[BMC, on the other hand, transformed the Austin-Healey from a car which was as fast as the TR3A, to one which left it well behind. 1958 'Big Healeys' had about 140 bhp, but by 1960 this had risen to 180 bhp, while the weight of the cars fell, and the chassis improved considerably.]

The team's performance in the Monte Carlo rally, with brand new red TR3As, was a great disappointment, for the roads were almost completely dry and clear on the 270 mile mountain circuit, there were so-called 'secret' checks which seemed to be anything but that to a few favoured French crews—and none of the drivers in the TR3As were French. Wallwork's car blew a cylinder head gasket, but kept going, while Annie Soisbault went off the road, as did Gatsonides at one stage.

The works team was looking for a boost in the Tulip rally, and was well rewarded when Keith Ballisat not only won his class, but finished second overall—the best-ever result in a

Johnny Wallwork took a brand-new 'works' TR3A on the Monte Carlo rally of 1959, but the car's agility was not enough, for razor-sharp time-keeping on the regular sections (with so-called 'secret' controls which the French seemed to know all about) were essential to success! (Johnny Wallwork)

Above Keith Ballisat (pointing) and Alan Bertaut drove WVC 250 in the Monte Carlo rally of 1959, but without success. (Keith Ballisat)

Above right Keith Ballisat's car control, at the wheel of a TR3A, was well known. He needed all of it here on the slippery surface of a mountain pass close to Monte Carlo in the 1959 event. (Keith Ballisat)

European Championship rally by a TR3A, and this in a 2-litre (not a 2.2-litre) car. For once the Tulip was a very tough event on the road, which suited the professionals as opposed to those who only entered it as part of a spring holiday. Only five cars finished unpenalized on the road, of which Keith Ballisat was one, while Johnny Wallwork's car lost a single minute and Gatsonides two minutes.

Cyril Corbishley, not even seen behind the wheel of a 'works' TR3A, rounds the Hunze Rug corner, behind the pits, during the last test of the Tulip rally of 1959. David Seigle Morris, in his own TR3A, is close behind him. (Cyril Corbishley)

Starting the last speed test of the 1959 Tulip rally—Zandvoort Grand Prix circuit, but in reverse—are three of the 'works' TR3As. WVC 247 was driven by Maurice Gatsonides (3rd in class), WVC 249 by Johnny Wallwork (2nd in class) and WVC 250 by Keith Ballisat (2nd overall, and the class winner). (Keith Ballisat)

This event was noticeable as the last in which Maurice Gatsonides drove for the team. His relations with Ken Richardson had become increasingly strained in the last year or so, not only over the Alpine crash, but over the way he always wanted to take delivery of his cars before everyone else, to make further minor changes—and over the fact that he and Richardson no longer got on.

Now it was time for a change of emphasis—from rally road to race track. The TR3S Le Mans car, long rumoured in Coventry, but not previously seen, was finally unveiled in April

Standard-Triumph surprised everyone by producing this new racing car—the TR3S—to compete in the 1959 Le Mans 24 Hour race. Although looking superficially like the standard car, the TR3S had a six-inch longer wheelbase, a glass-fibre body, and a brand new twin-cam engine.

The 'Sabrina' engine which powered Triumph's Le Mans cars from 1959 to 1961 was an eight-valve four-cylinder twin-cam of classic Jaguar/Coventry Climax-derived design.

1959, looking very similar to a normal TR3A, but being very different under the skin. It was a project which had taken up a great deal of the department's time in recent months.

This was the realization of a dream Alick Dick had been nurturing ever since three TR2s had competed so valiantly in the French 24-hour race in 1955. Dick, however, saw that there was no point in returning with standard cars, so he agreed to the design of a new engine which:

> . . . could be basically a racing unit, but I wanted it to be able to get Triumph the team prize at Le Mans, and I wanted it also to be producible, even in quite small numbers, for a top-line version of the new TR; it also had to be a 2-litre engine, of course.

Although work began on the new engine in 1956, it never went ahead as a top priority, the result being that it was 1958 before the unit was on the test bed, and 1959 before three race cars were completed. The design of these cars has been analysed on several previous occasions. The chassis had a six inch longer wheelbase than the normal TR3As, there were four-wheel disc brakes, an anti-roll bar was used on the front suspension, the 'Sabrina' twin-cam engine was a totally new 150 bhp 2-litre unit, while the body shell *looked* like that of a TR3A, but was made from glassfibre.

The cars used TR3A look-alike styling, though in truth they were very different in many ways. The six inches extra length of wheelbase was used because the twin-cam engine demanded a larger engine bay. This had a knock-on effect on the body panel shapes, where the front wings were a lot longer (and somewhat higher) than standard.

When you consider that these were racing cars, intended only for use on the smooth Le

The TR3S Le Mans cars of 1959 (and the TRS cars which followed them in 1960 and 1961) used these Girling disc brakes on the rear axle.

This head-on shot of one of the three TR3S cars of 1959 confirms that it had the same track width as the normal TR3A. The windscreen was cut down as far as the regulations allowed, but it was still not an aerodynamically-efficient race car.

The Jopp/Stoop TR3S was running well towards the end of the 1959 Le Mans race when it was forced to retire with engine failure. Dickie Stoop is driving at this point.

Mans track, they were disappointingly heavy, as pre-race scrutineering recorded an average weight of 2,025 lb. On a track like this, was a thick GRP body needed? Was it necessary to have a stiffened chassis frame? Worst of all, was it necessary to have cooling fans on the engines?

Keith Ballisat (a reserve driver for the 1959 team effort) made an interesting comment, which probably signals the first sign of conflict between Ken Richardson and his bosses:

> The factory policy was that it had to look like an ordinary TR. That's where the big clash came. Ken wanted to do an out-and-out racing car to beat the Porsches, but he wasn't allowed to do so. Obviously the shape affected its speed at Le Mans.

Three cars started the Le Mans classic, and it wasn't long before Ken Richardson's pessimism was borne out. The drag developed by the new cars was such that they could only reach 135 mph (the 1955 Le Mans cars had been timed at 120 mph), although the fastest

XHP 259 had already competed in the Tulip rally of 1959 (driven by Cyril Corbishley) before Standard-Triumph lent it to the Cambridge University Automobile Club to use in long-distance record attempts at the Monza circuit in July 1959. (BMIHT/Rover Group)

lap achieved was in 4 min. 46 sec. (105.3 mph).

Two cars retired when their fan blades broke off from the engine, damaging the cooling systems and wrecking the engines. The third car, its blades removed, was going well in seventh place, when after 22 heart-breaking hours its oil pump drive failed. Apart from the engines, the rest of the cars had been totally reliable.

This was the first *and* the last appearance of these cars, because Standard-Triumph had bigger and better ideas for 1960. The three cars were dismantled, the old bodies being thrown out on to a scrap heap, and the rolling chassis retained for 1960. What happened next is still the subject of conjecture, for at least one of the 1959 bodies is said to have appeared on another car built up by a private TR enthusiast, but using a push-rod engine.

The fact is that XHP 938/939/940 never raced again, and were replaced by not three, but four, new-shape cars for 1960 and 1961 based on the 'Zoom' style, which was currently proposed for use in a new TR4 sports car.

Almost immediately after the Le Mans race had been completed there was another big team effort in the Alpine. Gatsonides's entry (which was originally in the programme) was scratched, but four 2-litre cars tackled the French Alpine rally. Richardson, by this time, had realized that his TR3A's could no longer beat the Austin-Healeys in a straight fight, so he kept three of his cars (Soisbault had a 2.2-litre car) well clear of their class. Surprisingly, the Healeys struck trouble, the privately-prepared Bennett/Galliford 2.2-litre TR3A won the class—*and* finished ahead of all the factory cars.

It was typical of Annie Soisbault's luck (and lack of professionalism?) that when her car punctured she could not make the jack work, claiming that a vital component of this tool was missing. Unhappily, Keith Ballisat, who had now discovered the joy of pace notes, crashed his car on Mont Ventoux and broke his jaw. On the other hand, it was nice to see that large man from Bristol, 'Tiny' Lewis, win his class in a privately-registered Herald Coupé. This car had the third chassis number in the entire Herald series.

[Incidentally, this so-called 'private' car, registered TL5, seemed to spend a lot of time in the workshops at Allesley having preparation work done on it. Gordon Birtwistle, later to become No. 2 in the department in 1965, spent a lot of time as an apprentice working on this car, and on the TR3As.]

This was the enthusiastic Cambridge University Automobile Club crew who used a 'works' TR3A to take several endurance speed records at Monza in July 1959. It was later converted to left-hand-drive for Annie Soisbault to use in rallies in 1960. (BMIHT/Rover Group)

Jean-Jacques Thuner of Switzerland (in the dark suit) won the 1959 Geneva International rally outright in his own TR3A. His reward was a 'works' drive in the Monte Carlo rally of 1960. He later became a regular member of the team in the TR4 period. (Jean-Jacques Thuner)

Then, in July, Triumph dabbled in record breaking, by lending one of the 1959 team cars—XHP 259—to the Cambridge University Automobile Club. This enthusiastic organization had already worked wonders with a lightly-modified Austin A35, but on this occasion they were aiming a lot higher—to achieve 100 mph for seven days, on the banked circuit at Monza.

The engine let the side down after more than four days, but no fewer than eight Class E records were achieved for 102 mph-plus runs at one day, two days, and three days, plus records for four days, 2,000 miles, 5,000 miles, 5,000 kilometres and 10,000 kilometres. The

Very smart—this must be before the start of the Liege–Rome–Liege rally of 1959, not after the finish. Keith Ballisat (in smart blazer) and Alain Bertaut ready for the off. (Keith Ballisat)

run actually stopped after three days, but the old records were so vulnerable that the four-day record was also captured.

In the meantime, Allesley was also preparing for the Liege-Rome Liege marathon, using the same four 2.0-litre cars which had tackled the French Alpine two months earlier. In what had been a rather eventful year, not always to the greater glory of Standard-Triumph or its drivers, the cars performed very well indeed. Only 14 of the 97 starters managed to finish (and some of them were absolutely exhausted), so two-out-of-four for the Triumphs was a real achievement.

Annie Soisbault was having one of her better weeks, for she finished fourth overall in the GT category, beating everyone except three Porsches (and outrunning the Big Healeys!) Keith Ballisat had obviously recovered very well from his shunt on the Alpine (though he used a different car!), finishing close behind, sixth in category and second in class. Frenchman Claude Dubois crashed his car, while Rob Slotemaker was fastest of all the TR3As, until the last night in France when he took a wrong turning and ran out of time.

In and around all this activity, Annie Soisbault was fighting hard for the European Ladies' Rally Championship, sometimes entering events where no other 'works' Triumph put in an appearance, and at other times joining in a major assault. Having started the year in a new red left-hand-drive car (WVC 248), she used it in the Monte Carlo, Sestriere, Tulip, Acropolis, Alpine, Liege-Rome-Liege, German and RAC rallies—which made it the hardest-working of all these cars.

Even so, after the Liege and before the German rally there was no way that it could be pressed into service in the ten-day Tour de France event, so she was obliged to use her own, white, French-registered machine instead—but won the Ladies' Award anyway!

The team's entry in the German rally, of October 1959, was a real break with Triumph-tradition (the team had never before entered the event), while another was that John Sprinzel (with Stuart Turner) was invited to drive a 'works' TR3A for the first time. As the cars ran in an 'Over 1.6-litre class' it was almost inevitable that they should all have to give best to an Austin-Healey 3000, this actually being driven by Pat Moss and Ann Wisdom.

More success in the Liege–Rome–Liege rally of 1959. WVC 248 was driven by Triumph's ladies' crew—Annie Soisbault (in light shirt) and Renee Wagner, who finished in a remarkable fourth place, while WVC 247 was driven by Keith Ballisat (holding the flowers) and Alain Bertaut, who finished sixth overall. Yet another spot lamp arrangement (see previous pictures for proof) is being tried, and the indicators have been moved out on to the wings! (Keith Ballisat)

The 'works' TR3As were so fresh after the Liege–Rome–Liege of 1959 that they were re-prepared and entered for the German rally just two months later. Keith Ballisat kept his Liege car (WVC 247), and Annie Soisbault, too, retained her car (WVC 248). Along with John Sprinzel/Stuart Turner these three cars won the Manufacturers' Team Prize. (Keith Ballisat)

Even so, it was a solid performance, for three cars started, three finished, and they won yet another Manufacturers' Team Prize.

Writing in his hilarious book *Sleepless Knights*, John Sprinzel brought more insight into Standard-Triumph's very busy season, for on this event:

> Unfortunately the car had not been re-prepared after its long outing [on the Liege-Rome-Liege, driven by Slotemaker], and someone had filled the brake-fluid reservoir with normal oil. The result was somewhat chaotic, in that both clutch and brake seals rotted away in a very short time, and it was only through team-mate Tiny Lewis's help that we were able to have some sort of brakes, the clutch being unsaveable. However, Stuart, and I managed to get through the rally using second and second overdrive for all the tight bits. . . .

The last event of the season was the RAC rally, which had been moved from its traditional March date, and had been turned from a driving test spectacular into more of a hard, long-

Annie Soisbault and Valerie Domleo shared WVC 248 on the RAC rally of 1959. In this shot Annie is alone at the wheel as she tackles the 'Rest & Be Thankful' hill-climb test in Scotland. (BMIHT/Rover Group)

'Tiny' Lewis drove his own Herald Coupé (Registered TL 5) in the RAC Rally of 1959, but he would later go on to become a valued member of the 'works' team in 1960 and 1961.

distance, endurance event. For the very first time, Triumph entered a team of 948 cc Herald Coupés (for Peter Bolton, Keith Ballisat and Cyril Corbishley to drive), along with a TR3A for Annie Soisbault. Two privately-entered TR3As were added to Annie's car to make up a team entry.

It all started badly for the factory when Annie Soisbault's British co-driver, Valerie Domleo, delivered the French girl to the start of the first driving test, on the lower promenade in Blackpool, and told her she would meet up with her after the test. Annie completed the test, drove on, couldn't find Valerie, drove on even further . . . And was about 15 miles out of Blackpool before she realized that 'after the test' meant precisely that—after the Stop Line!

Valerie said:

I'd given her up, even before we had really started! But eventually she came back into Blackpool and drove up and down the road above the test, tooting and flashing her lights, looking for me. Obviously we were late at the first time control, and that cost us the Ladies' Prize.

It also cost Annie her chance of winning the European Ladies' Championship of 1959.

The RAC Rally of 1959 is always remembered for the snowstorm in Scotland which blocked the road to the Braemar control. 15 crews drove round the blockage to find the control, others tried to force the pass and got stuck, while a third faction assumed that the control would be cancelled, and missed it out completely. All the Triumph team cars suffered badly.

The result was chaos, and an unexpected 'top ten' (in which the author finished fifth overall—having navigated his driver around the blockage). In the provisional results, the 'works' team of TR3As (Soisbault, David Seigle-Morris and Eddie Hodson) and the 'works' team of Heralds took the team prizes.

However, if protests had been upheld (they were not) there would have been a change. As Peter Garnier wrote in his 'The Sport' column in *The Autocar*:

Incidentally, if the Braemar protests are upheld, the two teams merely swop places, Heralds taking first place and TR3s second. . . .

Ken Richardson (reclining) and his assistant, Marshall Dorr, trying out a new type of reclining seat for a Herald saloon. This coincided with preparation for the Monte Carlo rally of 1960—the Herald saloon was being built up for a private entry.

1960

Early in January the competitions department at Allesley was crammed full of cars being prepared for the Monte Carlo. Not only were five Herald Coupés being built, but there were also two of the TR3As, along with a Herald being prepared for a private owner to use.

    Like its predecessor, this particular Monte was expected to be won by regularity and strict timekeeping, rather than by a high performance car, so the Heralds were thought to be fast

According to the publicists, Annie Soisbault and Annie Spiers intended to take this cheetah with them on the Monte Carlo rally of 1960. Could they have been serious?

Cyril Corbishley (driving) and Peter Roberts in full flight in the Herald Coupé in the Monte Carlo rally of 1960—they finished a very creditable 25th overall.

enough for this event. The two TR3A effort was blunted by the absence of Maurice Gatsonides (the acrimonious split of 1959 had been final), but his old partner Marcel Becquart was provided with one of the cars, while Jean-Jacques Thuner (who had won the 1959 Geneva rally outright, in his own TR3A) used the ex-Soisbault machine.

Except that the Heralds carried a lot of cold weather gear (radiator blinds, snow shields underneath the radiators, two spare wheels and the latest Dunlop Duraband studded tyres) they were standard cars, while the TR3As were to their usual 'winter-specification' which

Ken Richardson's team always used stick-on registration numbers for the Herald Coupés. The Keith Ballisat/Stuart Turner car's number plate was almost obliterated at this point of the 1960 Monte Carlo rally.

Jean-Jacques Thuner's 1960 Monte Car was kitted out with twin spare wheels, one of them carried on a rack on the boot lid. (Jean-Jacques Thuner)

included the carrying of a spare wheel on the boot lid.

In spite of all these detail efforts, it wasn't a good rally for the 'works' team. Keith Ballisat's car was crashed, unfortunately by his co-driver Stuart Turner, as Keith recalls:

> We were using studded tyres, and had started from Paris. I was asleep at the time, and Stuart was driving. The accident was all down to the tyres and the road surface—there wasn't anyone else around at the time, and we ended up in a ditch, we went in sideways. I cracked a pelvis, and we both ended up in a French hospital, me with a lot of plaster on.

Rob Slotemaker's car hit a rock which had fallen into the road, which ruined his radiator,

Keith Ballisat and Stuart Turner took WVC 250 to fourth in class in the Tulip rally of 1960, an event in which all the 'works' cars were beaten by David Seigle-Morris's privately-prepared car, which was registered D20. (Keith Ballisat)

For the 1960 Le Mans race, Triumph produced four new-shape cars coded TRS, which used the same basic rolling chassis as the 1959 TR3S cars, but were shaped rather like the 'Zoom' prototypes then being developed in Engineering. The four cars were registered 926HP, 927HP, 928HP and 929HP.

Becquart crashed his TR3A, while Annie Soisbault must surely have been distracted by all the publicity surrounding her turning up at the start of the event with a pet cheetah on a lead, claiming that she was going to keep it in the car until she reached Monte Carlo.

As usual it was the long regularity section—Monte Carlo to Guillaumes and back, by way of passes like the Col du Turini, which sorted out the results. Three Heralds qualified to take part, but the best result was that achieved by Cyril Corbishley in the blue and white ex-Press car, which finished 25th. Cyril's reward was typical of the rather unstructured way

A study in faces after the Tulip rally of 1960. Rob Slotemaker was driving the left-hand-drive WVC 247, with co-driver Ron Crellin, and had finished second in his class to David Seigle-Morris's privately-prepared TR3A.

This was the simple facia of the 1960 TRS race cars, complete with a large rev-counter, and very bare trimming. The windscreen on this car has yet to be fitted.

The TRS Le Mans cars had a very large fuel tank mounted in the usual TR position, on top of the rear axle line, and a 'bottle' jack was mounted on board in case of the need to change a wheel out on the circuit. The glass-fibre body construction is clearly visible on the boot lid and wheel arch panels.

that Ken Richardson ran his team—he was never asked to drive for the 'works' team again.

On the Tulip rally, which was a long and dreary run down to Monte Carlo, and back, with a few special tests and stages thrown in to alleviate the boredom, Triumph entered no fewer than four 2-litre TR3As (the 2.2-litre cars normally had to compete against the Austin-Healey 3000s, which was a hopeless task), and a Herald Coupé for 'Tiny' Lewis. The Herald won its class, ahead of Geoff Mabbs's own Herald Coupé (the positions would be reversed in 1961!), but all the 'works' TRs were beaten by David Seigle-Morris and Vic Elford, in a privately-prepared TR3A. If David had not lost his clean sheet on the ascent of the Col du Turini (only ten cars were unpenalized) he would have finished sixth overall in the rally—as it was, he had to settle for 19th place.

Somehow there had been a lack of 'spark' in the team's efforts on this event, for they were instead being drawn inexorably towards the Le Mans race. For its second assault on Le Mans with twin-cam engined cars, Triumph had produced four new cars which were called TRS. Mechanically these were basically the same as the 1959 TR3S models (and used almost all their running gear), but they were equipped with glass-fibre bodies of an entirely new shape.

Magazine pundits were scornful of the new shapes, pointing out that they weren't at all suitable for racing. They were right, of course, but then they did not realize that this was a shape currently proposed as the basis for the *next* generation of TR sports car—the TR4. In fact the 'Zoom' prototypes—for this, indeed, was the basic shape of the TRS—were about to be abandoned in favour of a shorter-wheelbase, revised version called 'Zest', but even that car would retain the same basic proportions.

Although the fan blade fiasco had now been tucked away, the TRS was no advance on the TR3S. The new cars weighed 2,180 lb.—which was 155 lb. *more* than in 1959, and the new shape was clearly no more efficient than before as the recorded top speed was 128.6 mph. In the race itself the cars lapped 10 seconds slower than the TR3Ss had ever done, although all three cars kept running to the finish. The best of the bunch (driven by Keith

The TR3S and TRS 'Sabrina' twin-cam engines were built on the 'sandwich' principle, with separate castings for the sump pan, sump, crankcase, cylinder block and cylinder head. The race cars always used twin-choke SU carburettors, though the still-born Conrero project would have used dual-choke Webers instead. This engine, at least, has no evidence of fan blades being fitted.

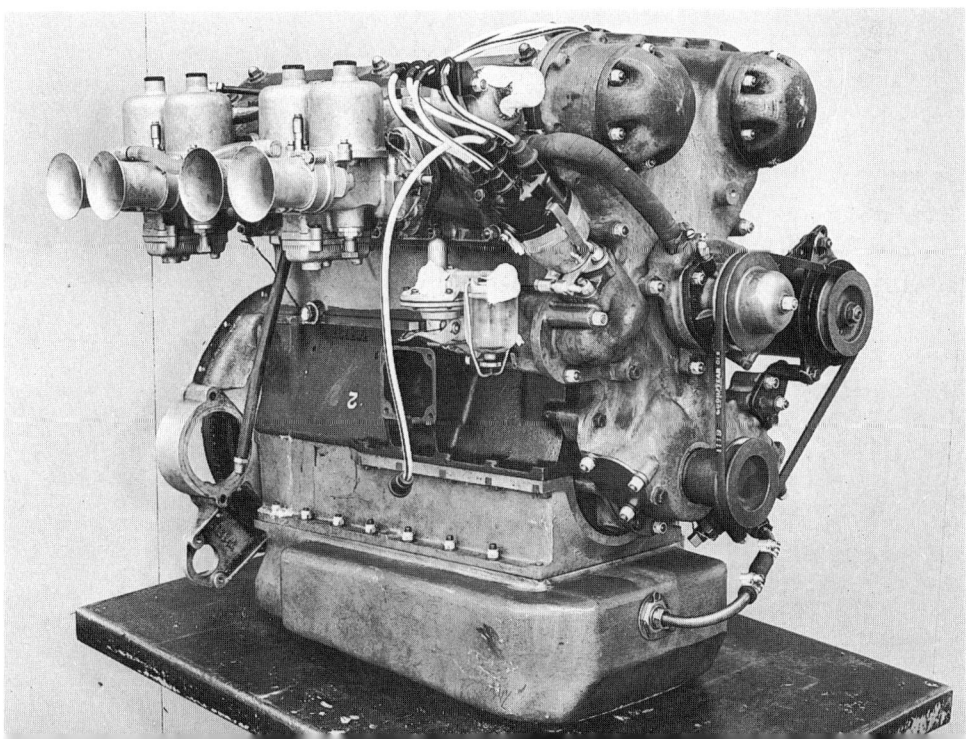

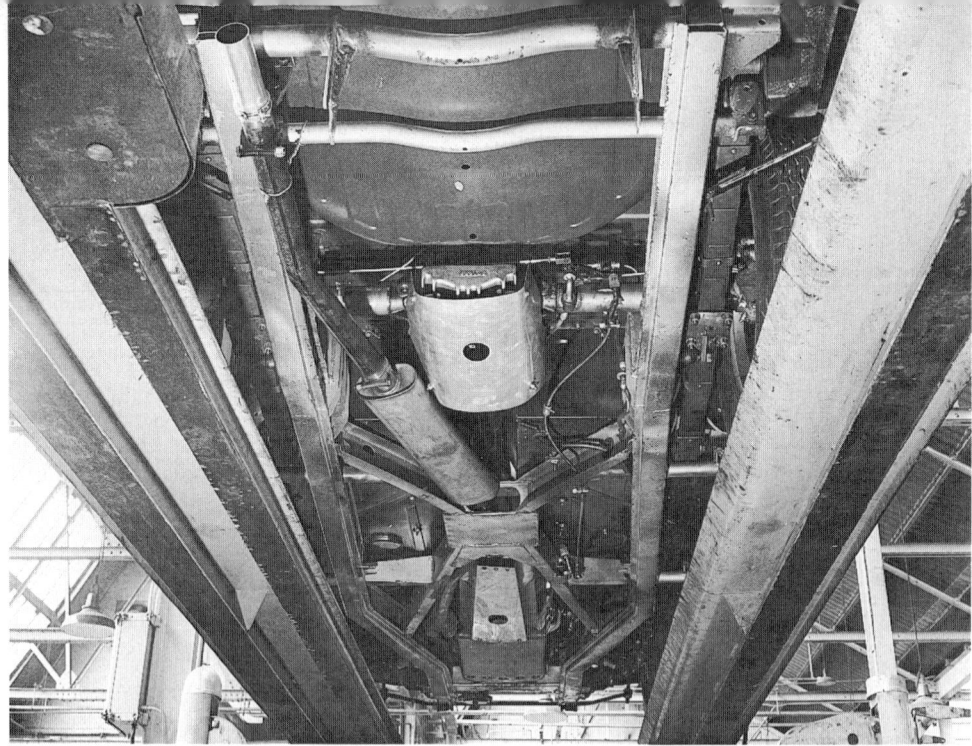

Underside view of the TRS Le Mans car of 1960, showing that the chassis side rails had extra-depth stiffening channels, and what looks like a cooling duct around the rear axle differential.

Ballisat and Marcel Becquart) averaged 89.56 mph and finished 15th overall.

The engines of all three cars suffered from valve seat insert 'hammering' during the event, which meant that valve clearances were disturbed, and power was lost. All three cars, however, made it to the finish, and there was optimism about the cars' potential for the 1961 race.

The Sabrina twin-cam engine installed in the TRS, showing that, except in detail, this could have been the prototype for a production car.

The Leston-Rothschild TRS makes a pit stop in the Le Mans 24 Hour race of 1960. Mike Rothschild is behind the wheel. This car finished well, averaging 88 mph.

A few days later a full team of TR3As took part in the Alpine rally, an event which only served to show how far the deadly rivals—the Austin-Healey 3000s—had advanced, and how TR3A rally car development had stagnated. Whereas the 3000s finished second, fourth and sixth overall, the best of the TR3As—that driven by 'new boy' David Seigle-Morris—finished ninth.

Les Leston at the wheel of one of the 1960 TRS cars which contested the Le Mans 24 Hour race in 1960.

Le Mans 1960, and the three TRS twin-cams line up at the pits, nearly four hours before the start of the race. 929 HP, the 'spare car' was not used on this occasion, but got its chance to shine in 1961, when 928 HP stood down.

All the TR3As ran in 2.0-litre form, though Rob Slotemaker's car broke its rear axle almost immediately after the start, Les Leston went off a few hours later, and Annie Soisbault's car lost a wheel on a tight section in Italy. Even though Seigle-Morris's car won its class, compared with the great days of 1956 and 1958, this was a real let-down.

By this time, in the summer of 1960, Standard-Triumph's business was beginning to fall apart, for sales were dropping, and healthy profits had turned into soaring losses. By November the situation was desperate, so when Leyland Motors announced a takeover bid in December 1960, Alick Dick was greatly relieved.

In the meantime, though, operating budgets had been slashed in many departments, and Richardson's motorsport department could not escape this. After the Alpine rally of 1960 the department was effectively put into hibernation, with almost all the Heralds, and all the

Tiny Lewis and Tony Nash used this Herald Coupé (YRW 269) to compete in the Alpine rally of 1960. The front tyre on the outside of the car is almost ready to roll off its rim.

Geoff Mabbs and Leslie Griffiths in Geoff's own Herald Coupé, which won the Tulip rally outright in 1961 when Tiny Lewis withdrew the 'works' car to give Geoff the best possible 'class advantage' position in the handicap system.

TR3As, being sold off. One of the TRs—WVC 248—was bought by ex-apprentice Gordon Birtwistle, who re-entered the Triumph motorsport history a few years later.

At this time, too, the department was moved out of the Allesley service department, and for the next (very restricted) year of its life it operated from a workshop in the Capmartin Road factory in north Coventry, a facility where Standard-Triumph's transmissions and axles were produced.

1961

Early in 1961 Ken Richardson must have been miserable, for his 'works' motorsport programme had been slashed to ribbons by fierce budget cuts. In 1959 his team had tackled 11 events and the Le Mans race—in 1961 there would be just three team entries—single cars in the Tulip and Acropolis rallies, and a well-funded return to the Le Mans race in the TRS twin-cams. His team of drivers had been dispersed—some, like David Seigle-Morris and Stuart Turner, going on to greater things with other teams.

The first 'works' entry of 1961, therefore, was for 'Tiny' Lewis to drive a Herald Coupé in the Tulip rally. Alongside Lewis, among seven other Heralds, was a privately-prepared Herald Coupé, which was to be driven by his close neighbour in the Bristol area, Geoff Mabbs. As in previous years the Tulip, which started and finished at Noordwijk in Holland, was run on a handicap basis, but this time it was fiendishly complicated and would ensure that the most outstanding car *in its class* would also win the event outright.

131 crews started, of which only 19 retired before the end of the event. It was a long event—the overnight rest halt was Monte Carlo!—but the meat of the competition was in a series of 14 speed hill-climbs and special stages, which were supposedly secret, but for which a surprising number of crews had practised, and had pace notes!

At half-distance, in Monte Carlo, although Mabbs's privately-prepared Herald was actually beating the 'works' car, there was no Triumph in the top ten positions. On the way back the two Heralds maintained their class position, but on the long run to Noordwijk Lewis and his co-driver David Stone made a series of hasty calculations. Mabbs and himself were a long way ahead of the next car in the class (an Auto-Union 1000S). To quote *The Motor:*

> Due to a complicated marking system . . . the Mabbs/Griffiths victory was only assured when

Ken Richardson savours the last outright win by a Triumph in the time when he was competitions manager, while Geoff Mabbs wonders how much to ask him for a bonus after the Tulip rally success of 1961!

'Tiny' Lewis retired his similarly-classed Herald, literally at the front gates of the Huis ter Duin Hotel, the rally headquarters and finishing point. . . . Lewis sacrificed an assured second place in his class—and quite a high place in the overall results—for a chance of an outright Herald win. The gamble paid off.

Standard-Triumph was ecstatic—and promptly put the car on display at Canley on its return, but there was little reward for Mabbs or Lewis, as only one event in the rally programme remained—a single entry for the genial Bristolian in the hot and dusty Acropolis rally, using a TR3A (WVC 251) which had never previously appeared on an international rally. WVC 251 had been Richardson's own road car for some time.

After the Tulip, however, the Acropolis rally was something of a disappointment. The TR3A was outpaced by solid saloons like Eric Carlsson's Saab, and the factory-entered Sunbeam Rapiers, so 'Tiny' settled for tenth overall, and second in his class to the Austin-Healey 3000 of Peter Riley. If only I had known, this was a portent for the 1962 and 1963 seasons.

But that was the end of the rallying effort in the Richardson era; the only event which now remained was the Le Mans race of 1961. Unlike the rest of the programme, it was a well-funded effort—almost a gesture of defiance to the accountants—and at the third time of asking it was to be a great success.

The lessons of 1960 were that the engines needed more power and more reliability, and that they needed to handle better. With very few other events to get in the way in the winter months, the quartet of TRSs was thoroughly reworked, and appeared at the April 1961 Le Mans test days with wider tracks, rack and pinion steering, with extra flaring of the wheelarches to cover the repositioned wheels, and with hot-air exits from the engine bay in the side of the front wings. The wide-track front suspension and the rack and pinion steering were both put into production, in the TR4, only a matter of months later.

Although they were no lighter than before, and kept their very unaerodynamic styles, the latest cars were quicker in practice, and a lot quicker in the race itself, where the 155 bhp

After Geoff Mabbs won the Tulip rally of 1961 in a privately-prepared Triumph Herald Coupé, his car was put on show outside the main offices of Standard-Triumph. The placard tells its own story.

engines kept their tune to the end. Lap times were reduced by a full eight seconds, the fastest lap was at more than 105 mph, and over the 24 hours distance the best of the cars averaged 98.91 mph, and finished ninth overall.

Keith Ballisat and Peter Bolton shared that car, but the two other cars finished 11th and 15th and—finally, but gloriously—Triumph won the Team Prize. Significantly, the Austin-Healeys all retired, and the Sunbeam Alpines were well beaten.

Triumph's team of four TRS cars arrives outside the team's hotel, before the start of the 1961 Le Mans 24 Hours race.

Rob Slotemaker (scratching his head) and Les Leston (whose back is giving him trouble), with their car before the start of the 1961 Le Mans 24 Hours race. They finished 11th overall.

As *Autocar's* Harry Mundy summarized afterwards:

> The only team to finish the race intact was Triumph—an achievement indeed . . . the race average of their fastest car showed an improvement of the order of 10 mph.

There was only one significant problem which delayed one of the cars. On the Becquart/Rothschild machine (929HP) a camshaft oil seal broke up and started leaking oil on

926 HP, driven by Keith Ballisat and Peter Bolton, finished ninth overall at Le Mans in 1961, but at this instant it is about to be swallowed up by a Ferrari 250GT! The other two TRS cars are in the rear of this high-speed traffic jam.

to the exhaust system. As oil could only be added at specified (long!) intervals, there came a time when the car had to be halted at its pit to wait for the end of the race. Close to the traditional 4 p.m. finish it was sent out once again, and took its place for a three-abreast 'formation finish' which was caught on film. The delay explains why the first and third Triumphs were separated by no less than 183 miles (more than 90 minutes' racing) at the finish.

Ted Eves of *Autocar* took a test run of the Bolton-Ballisat car after the event—on the country roads of Warwickshire, complete with competition numbers and a loud exhaust system. He also inflicted Coventry city centre traffic on the car and later wrote that:

> I was duly thankful that there was no tendency to oil a plug at low engine speeds, for at 3,500 rpm torque *and* noise came in together with embarrassing *éclat* . . . it was quite possible to dawdle at 2,800 rpm in top.
>
> Once the road was derestricted a change down into second sent the revs rocketing to 6,500 almost before I could stop them; 6,500 in second corresponds to 76 mph and it took just six seconds from the 30 limit. A change up into third, and 100 was reached in 17 seconds from the derestriction sign. . . .

When it was being designed, that engine had also been intended for use in a road car, with single-choke SUs and a 120/125 bhp power output. If the company's finances had been healthier this might have come to fruition, but by 1962 all had been forgotten, and the project was cancelled.

There was one other 'might-have-been' twist to the race car programme. During the autumn of 1960, and well before preparation of cars for the 1961 Le Mans race had been considered, the directors had agreed on the development of the next generation of race cars. The board minute states that: 'The Italian tuning expert, Conrero, is to build up four new Le Mans cars for the 1961 race at the approximate cost of £25,000.'

As it happened, the company's cash crisis soon played havoc with the motorsport budget, and the result was that only one car was ever finished, and even that car was not returned to Coventry until the 1961 race had been run.

But what a car! Always known as 'the Conrero' at Standard-Triumph, it combined all the TRS's normal Le Mans running gear in a tubular chassis-frame, which was topped off by an attractive lightweight light-alloy closed two-seater body shell (in red) and right-hand-drive. Styling was by Michelotti (with body construction by his long-time associates, Vignale), and this was the only TRS-based car to have a Sabrina engine fitted with two dual-choke Weber carburettors.

Gordon Birtwistle, who had joined the Engineering department by that time, remembers the Conrero car with affection:

The TRS Conrero was completed too late to take part in any races before Leyland closed down the motorsport operation in 1961. With this beautiful Michelotti-styled coupé body it certainly looked the part.

Three-quarter rear view of the 1961 TRS 'Conrero' car, which Triumph might have used in the 1962 Le Mans race if the department had not been closed down. Unlike the 'works' TRS cars, this one had a metal body shell.

This was definitely a Harry Webster project; Ken Richardson had nothing to do with it—in fact I don't think he ever saw it. It had an entirely different frame from that of the TRS, it was definitely not a standard TRS frame.

It was a wonderful car, a lot faster in a straight line than the TRS. I tested all the Sabrina-engined cars before they left for the USA. The Conrero was better than the open-topped 'flying rectangles', which were truly agricultural machines. When you drove one of those you certainly knew you'd been driving one! *Everything* was heavy—especially the clutch.

The Conrero, though, was much quicker, because it was a GT coupé, and more aerodynamic.

The one-off TRS Conrero prototype looked purposeful except, for goodness sake, for having an umbrella-handle handbrake!

In the TRS Conrero the Sabrina twin-cam engine was fitted with two twin-choke Weber carburettors, not twin-choke SUs as used in the 'works' race cars.

It was so much more refined, and quicker, I would say, because it didn't weigh any more and had a better aerodynamic shape.

Incidentally, Harry Webster drove it as a road car for a short time!

The Conrero car certainly looked the part, and was probably good for more than 140 mph, although it was never tested on a track by a racing driver, so we can only guess at its potential. It was registered 3097 VC (a Coventry registration which dates from the spring of 1962) before being sent to the USA later that year. By the 1990s, and after many years of neglect and storage, it returned to the UK.

Closedown in 1961

Leyland's action following the Le Mans success was swift and brutal. The four TRS cars were removed from the competitions department at Radford, delivered to the engineering department at Fletchamstead North, and stripped out for inspection. Competitions was then closed down, Ken Richardson was made redundant, and the team of mechanics was dispersed. Within weeks there was no trace of former glory. Richardson, however, recently said that the Leyland management changed its mind, and asked him to return—but that he refused the offer.

Towards the end, no doubt, there had been a great deal of friction between Ken Richardson and his bosses in Engineering, which must have led to the break. Perhaps Richardson himself sums up the stubbornness for which he became known: 'What I wanted done I had done, and nobody could change it.' That might have been acceptable ten years earlier, but by 1961 those days had gone.

This is the point at which Ken Richardson fades out of the Triumph story, almost as suddenly as he arrived. In the following year, 1962, he ran a restricted motorsport effort for TVR in Blackpool, but this was also cancelled when that tiny company ran into financial difficulty. Ken then ran his own garage business for a time, but eventually returned to Bourne in Lincolnshire, and to retirement.

What of the twin-cam engined cars? All five of them were eventually rebuilt, then sold off in the USA. Many years later, in the 1980s, most of them reappeared on the 'classic car' market, and although spare parts for the twin-cam engines were no longer available they were restored and run once again.

But that wasn't the end of Standard-Triumph's motorsport activities—as the author was about to find out.

Chapter 4

Under Leyland control
New management, 1962 and 1963

Although Ken Richardson's department was closed down abruptly in 1961, Standard-Triumph did not stay out of motorsport for long. Starting in February 1962 a new department was opened up, managed by the author, but this time firmly under the control of Harry Webster and the Engineering Department. For this reason, I hope the reader will allow me to make this, and Chapter 5, rather more autobiographical!

Almost as soon as the old Competitions Department had been closed down in the summer of 1961, Standard-Triumph's management made the first moves to start up again; it would have been impossible to bend the old regime to its wishes, but a new start was more straightforward.

The sales force, in particular those managers who were the export specialists, was convinced that a continuing motorsport programme was essential to bolster up Triumph's image. The launch of the next-generation TR sports car—the TR4—was imminent, and development of a new small sports car, the Spitfire, was well advanced.

At this point, in my own small way, I enter the story. I had been living and working in the Midlands since 1957, had started rallying in 1958, had concentrated on co-driving, and had already received offers from the BMC rally team in the first few months of 1961. Then, in April of that year, I joined Standard-Triumph as a development engineer, and the offer immediately lapsed!

It is also worth noting that Coventry's fastest-growing motor club of the period was the Godiva Car Club. Naturally I was a member, so was Harry Webster of Standard-Triumph (who liked to go rallying himself on local events), and so was Lewis Garrad of the Rootes 'works' rally team.

Later that year Lewis invited me to join the Rootes 'works' rally team, as Peter Procter's co-driver who was driving the successful and well-developed Sunbeam Rapiers. Both my immediate boss at Standard-Triumph, John Lloyd, and *his* superior, Harry Webster, approved of this, but there were always two provisos—one was that I couldn't take time off to go rallying, but had to use holiday entitlement instead, the other was that there should never be a clash of interests, and no adverse publicity!

The local Press was very good about this. As far as I was concerned, it was a good deal, which Peter and I celebrated by winning a British national rally in October, by being members of the team which won the Manufacturers' Team Prize in the 1961 RAC rally in November—and by finishing fourth in the Monte Carlo rally of 1962, and helping to win the team prize once again.

Even before then, however, Harry Webster had shown me a memo from John Carpenter and Lyndon Mills of the Sales department, who argued that Standard-Triumph needed to be in motorsport somehow, if only by making an annual entry in the Alpine rally. Harry, who had already spent many social hours with me at the Godiva Car Club, and who always made his love of Le Mans very clear, wanted to know what I thought?

Having got involved with 'works' teams by this time, and having seen the way that BMC, Rootes and Ford were all becoming much more professional, with more specialized cars, I had my own ideas. Perhaps I was cocky, but I was convinced that a reborn competitions department should be very different from the old, and have different policies.

In a long discussion paper about Standard-Triumph's potential—and its future—I dismissed the idea of a one-off entry in the Alpine (it would have been impossible to find competent drivers on a one-off basis), and detailed a choice of strategies:

> A major programme to match the opposition—which effectively meant BMC and Rootes. This would be costly, and would take two or three years to mature.
>
> A limited programme, using TR4s, in rallies.
>
> Abandoning motorsport completely, on the grounds that success was never assured, and that *any* effort was likely to be very costly.

For a time there was a complete silence, though as I wasn't sacked I reasoned that I couldn't actually have offended too many people. Nothing happened until January 1962, by which time Harry Webster had made up his own mind. Standard-Triumph had a new managing director, Stanley Markland, who was a bluff Lancastrian from Leyland Motors, already with a formidable management record in that company. He had listened carefully to the arguments, and agreed to a limited programme, but he wanted Harry Webster to control the department.

Harry called me in to his office one day, spelt out what he had in mind, said it was time that I stopped indulging in a hobby with a rival firm, and that he wanted me to become his Competitions Secretary. You could say that this rather upset my vision of the future, for I was looking forward to more drives with Rootes in 1962 and—who knows?

Instead I had to make the big decision—should I take Harry Webster's offer, or should I leave Standard-Triumph, try to find a 'day job' to keep me alive between events with Rootes (for rally drivers and their partners made no money from the sport in those days!), and carry on rallying?

In the end I took up Harry's offer, and even before the Monte Carlo rally of 1962 which Peter Procter and I so nearly won (he slid off the road on one special stage, which cost a wheel change and about four minutes), I had told Norman and Lewis Garrad that I would have to leave them.

The new Competitions Department *was* new, in all respects, for it would have new personnel, new cars, new management control, and it would operate in a new workshop. Although I was allowed to look at the entire product range (current and soon due for launch), it was clear that if the new team was to be up and running within months there was only one potentially competitive car—the new TR4. In February 1962 a small team of mechanics, led by Ray Henderson (who had a lot of experience in the old TRS Le Mans race car programme, and had led the intrepid mechanics across Africa with the Heralds in 1958, and who later took over from me in 1965) began preparing to build cars.

But where should that be? There was never any question of using the gloomy old 'Le Mans' shop at Radford, which had already been converted for transmission test and development purposes. Fortunately for all concerned, Leyland had just cancelled one of Alick Dick's last pet projects, the development of a tractor to compete with the Ferguson interests which he had only recently sold off, which vacated a large and comfortable workshop in the main Engineering and Development building at Fletchamstead North. This was allocated to 'Comps', where John Lloyd and Harry Webster could keep an eye on it.

As the new deal involved 'borrowing' expertise from the engine build and test departments, and from the body development department, all of which were under the control of John Lloyd, this was an ideal set-up. However, it was a tiny operation—I remember that the budget for 1962 was £12,000, but I also remember that we had to fudge all manner of things to get work 'hidden' in other project budgets—and as Competitions Secretary, I had no assistants, and no secretarial help. This was the period in which my typing skills went ahead by leaps and bounds.

In February 1962, when we started work, one really major problem overwhelmed all others—there were no records, and very little expertise, to draw on. Seven months earlier, when Ken Richardson had been made redundant, he had taken away, or destroyed, every scrap of archive material and preparation data. To learn from the old TR3As and Heralds, which had all been sold anyway, we even had to buy new copies of the homologation forms from the RAC!

Three factors helped us decide that we would have to start from scratch once again—one was that the TR3A homologation forms were not at all informative, another was that Gordon Birtwistle (who had bought the ex-Annie Soisbault TR3A, WVC 248) told me that it was no faster or better-handling than an ordinary TR3A, and the third was that Standard-Triumph's engineering departments seemed to have had no regular contact with the Richardson-run department for some time.

When the time came to 'make the rounds' of the engineers, I was amazed to learn that no higher-power tune-up work had ever been done on TR3A push-rod engines by the factory's engineers. Indeed, there was a die-hard faction who had once seen crankshaft fatigue failures at 5,200 rpm, so they stubbornly suggested that engines should never be revved over 5,000 rpm, and should never be modified.

The fact was that specialists like SAH Accessories and Lawrencetune knew a lot more about the wet-liner engine than Standard-Triumph did, and drivers had habitually been taking them to 6,000 rpm and more for years without seeing any ill-effects. Although the politics of the situation were unfavourable, and funds virtually non-existent, I would have to consult these people—soon.

Even before we took delivery of four new TR4s, never mind starting to prepare and develop them, Ray Henderson's team was asked to prepare a Herald 1200 for, of all things, the East African Safari rally. An American driver, Bob Halmi, who had previous driven TR3As in European events, had asked for it to be built—and the money was good.

In those days this event was for Group 1 cars, nearly standard, and the East African roads were rougher than they are today, so it was asking a lot for a Herald to survive, let alone be competitive. With no preparation data available, all we could do was to cheat like mad when it came to chassis stiffening, study all the *pavé* test sheets to see where the body also needed some illicit reinforcement, and hope for the best. It wasn't enough. The car broke down, and this was our first and last involvement with the Herald.

In complete contrast to the policies of the old department, we fixed a programme and a strategy at the start of the year. We decided to tackle the Tulip, the French Alpine, the Liege-Sofia-Liege and the RAC rallies in 1962, leaving a bit of flexibility for a single entry in other events. We also acknowledged that the performance of the latest 'works' Austin-Healey 3000s (complete with Weber carburettors and 210 bhp) was way out of our reach, so we resolved to keep out of their classes wherever possible. As with the old TR3A, both the 2-litre and 2.2-litre engines were homologated.

Harry Webster wanted to run three TR4s in all events (the Team Prize was always a desirable trophy to be won), but finding good drivers wasn't easy, especially as rival teams tended to tie up drivers for the season, and by February 1962 all the deals had already been done.

Nor was there any advantage in looking back into the Richardson era. All the good young drivers who had been 'discovered' by Ken Richardson in the 1950s had moved on to different teams, and others—like Wallwork, Waddington and Goldbourn—were growing old. Annie Soisbault had married a rich Frenchman and turned to racing, while Keith Ballisat's Shell connections had drawn him to the Rootes Group. We had to start anew.

In the short term I turned to John Sprinzel, who had made his name in Austin-Healey Sprites, but was known to be a maverick, a loner. John joined the team, and since he thought that we needed: 'One old lag, one overseas driver, and one s**t-or-bust merchant', I also signed up Mike Sutcliffe (who had recently being driving for Ford) and

Jean-Jacques Thuner of Switzerland, who was really a pace-notes specialist. Roy Fidler, who would later become British rally champion in his own right, joined Sutcliffe as co-driver.

Compared with BMC and Ford, we didn't have world-class talents, but it was a start. Johnny Wallwork was offended that he wasn't asked—but he *was* 49 years old at the time—and at that stage I couldn't invite Rob Slotemaker to drive as I was only 'allowed' one overseas driver.

The TR4s arrived in February straight from the production lines at Canley, and we even managed to get some easily remembered registration numbers—3 VC, 4 VC, 5 VC and 6 VC; 1 VC and 2 VC were not available from the authorities in Coventry by the time we asked. Because all 1960s 'works' cars tended to run in recognizable 'team colours' we chose powder blue hardtop cars with 2-litre engines, overdrive and wire wheels. The reason for the colour choice was that BMC had a monopoly of red, British Racing Green didn't photograph well, and white got dirty too quickly!

If we had settled for the Richardson formula of 'standard cars with rock hard suspension', preparation would have been simple. Advice from the drivers, however, was that the old cars had been much too hard—and that we needed to make the TR4s faster. The first job, accomplished with the help of Fred Nicklin and Gordon Birtwistle (experimental test drivers) was to make the cars handle well without being too hard, which was speedily done by choosing Koni dampers for the front end, and by fitting larger Armstrong dampers at the rear.

Right from the start we decided to used brand new chassis frames for every event, reinforcing them around the front and rear suspension fixing points, and to paint them white so that we could pick up any cracks and creases as they appeared. My bitter experience in other cars, and other rallies, also led us to re-route the brake pipes, and to armour them, either with rubber sheathing or (in the case of flexible pipes) with spiral wire.

BP provided sponsorship—and free fuel coupons, Dunlop provided free tyres, Lucas sent technicians over from Birmingham to build completely new wiring harnesses, and we managed to reserve a couple of experimental department 'hacks' to use as service cars. It was a start, but it wasn't as glossy or as well-funded as legend might suggest.

Right away, too, Ray Henderson and I studied the homologation regulations very carefully, took our best caps in hand, and went out shopping for help from the engineers. Pleas for higher output engines met with blank faces (and a refusal to spend test bed time making up for eight years of neglect), while no-one had ever tried to fit a limited-slip differential to the rear axle, but at least we got assurances that the Vanguard axle ratios (4.3:1 and 4.55:1) would fit, and we really struck gold among production engineers who said that, yes, they *could* produce light-alloy body skin panels for us, if we waited until the end of each run of stampings for a particular part.

John Sprinzel and I took the first completed car—6 VC—on the British Trio rally to make sure that nothing would fall off—nothing did—before the other three cars tackled their first event, the Tulip rally of 1962. We were all rather wet behind the ears, but at least we had taken weight out of the cars with the aluminium panels, by fitting plastic instead of glass side and rear windows, and by removing the bumpers.

We had also been able to recce all the hill-climbs (John Sprinzel and I used 6 VC for this job); even though the route was theoretically secret, BP-France discovered the road closing orders from the French police—as did every other sponsor or organization in that country!

The problem was that we knew the 2-litre engines lacked power (for the first event only, we had to run in standard condition, with a 5,000 rpm limit, though 4.55:1 axles were fitted), and we *all* lacked experience. So, perhaps it was no wonder that, although the three cars finished well, they were beaten by a 'works' MGA driven by Rauno Aaltonen. It was the first and last time I tried to combine co-driving with team management—there was simply too much to do for both jobs to be tackled efficiently.

The difference between the TR4s and their rivals was small but obvious. The MGA was

What, three-up in a TR4? This was Mike Sutcliffe's car, 4 VC, at the finish of the 1962 Tulip rally, where it had finished third in class. Co-driver Roy Fidler is driving, Mike is in the passenger seat, and that is Mike's wife Jackie on his knee. (Roy Fidler)

homologated to its limits—with four-wheel disc brakes, Weber carburated engine and all. *That* was what we would have to do if the TR4 was to be competitive.

With the Alpine rally in only four weeks time—and this was the traditional event where a TR was expected to shine—everyone got stuck in to making the cars faster. Even more weight was taken out of the cars, by stripping out the interiors, and running without tool kits or spares. More important, work started on the engines.

On the Alpine, where the class limit was 2.5-litres, we converted the engines to 2.2-litres, and fitted SAH heads, newly homologated SAH tubular exhaust manifolds and a high-lift camshaft. On my own responsibility, I also allowed a rev-limit of 5,500 rpm—but afterwards I found that the drivers were all using at least 6,000 rpm in any case. From memory, I recall that the improvement in power/weight ratio was something like 22 per cent—

Mike Sutcliffe's class-winning TR4 on the Passo di Vivione, in the Italian Dolomites, in the 1962 Alpine rally.

and the cars certainly felt (and were) a lot faster. On this occasion, where there was a long speed trial at Monza to be considered, a 4.3:1 axle ratio was fitted.

This was the event where all four blue TR4s were used—6 VC not only completed a recce but was driven on the event by Tommy Wisdom and Jeff Uren (Tommy was still an important journalist, and Public Relations were paying his bills) with a really outstanding result being achieved. Sprinzel's car was shunted out of the event by a non-competing car on a road section, but the other three cars kept going, and with less than four hours to go two drivers (Sutcliffe and Thuner) were still unpenalized.

Although the 9,000 feet Stelvio pass was still blocked by snow, all the other traditional horrors were open—and had been used—including the Allos, the Vivione and the Cayolle, but the hard work was all due to end with two extremely tight final road sections in the mountains behind Nice. In those days the two sections—Entrevaux to Sigale, followed by Sigale to the Quartre Chemins cross-roads—were timed at 60 kph, and were only *just* cleanable if the crew had practised, if the roads were dry and if nothing went wrong.

Roy Fidler, tackling his first Alpine, was as nervous as a kitten, but remembers the advice he got from Sprinzel's co-driver, Willy Cave:

> Both sections were at least 35 km long. You had to go like a bat out of hell on the first section and know the time you needed before putting the card into the printing clock. As I jumped out I remember Willy standing there and screaming: 'What time do you want?' I remember telling him, and he shouted back: 'Don't stamp yet, your time hasn't come up.'
>
> I stood there, trembling, wondering if I'd got my arithmetic right. If It was the most horrendous minute or so, because I knew that Mike would kill me if I'd got it wrong! Then we set off, exactly at the right second, but we were still well into our last minute at the end of the next section. So we'd kept our *Coupe*, and after that it was easy.

Jean-Jacques Thuner, who often seemed to have awful luck in his time with Triumph, also set off, unpenalized, on the second section, but had to stop and change a punctured tyre. Or so he thought. In fact he had clipped a rock, damaged the steering, stopped, leapt out to change the wheel, then found that there was no puncture after all . . . but it was all too late, and he missed his *Coupe* by less than a minute.

Even so, it was good to see Sutcliffe finish fourth overall, Thuner sixth, and the three surviving cars take first, second and third in their class. If only BMC's Austin Healeys hadn't

Mike Sutcliffe's Coupe-winning TR4, complete with 2.2-litre engine, on the gravel surface of an Italian Col in the Alpine rally of 1962.

The 'works' TR4s took first, second and third in their class in the Alpine rally of 1962, thus carrying on a great Triumph 'tradition' in this event. Four cars started, and these three finished. Left to right: Jean-Jacques Thuner, John Gretener, Mike Sutcliffe, Roy Fidler, Tommy Wisdom, Jeff Uren. (Roy Fidler)

performed so well, Triumph would have won the team prize too. This was the first time, too, that I had been able to use a Vitesse, as a high-speed 'chase' car.

Then came the Marathon—the Liege—an event which had got rougher, faster, and even tougher since the TR3As had last tackled it in 1960. The turn-round point was now in Sofia, the run back was all the way along the dreadful tracks of the Adriatic coast of Yugoslavia, and the schedule was more demanding than ever before, with higher average speeds, and earlier control closing times.

Nothing to be proud of here, for Mike Sutcliffe had crashed 4 VC on the Liege–Sofia–Liege rally of 1962, but it shows off the construction of the 'works' TR4 bodies, which were running with engine bay vents for the very first time. (Gordon Birtwistle)

More than any other event of the year, this convinced me that the TR4 was no longer a good prospect for rough events, as it was likely to be battered to destruction by the rocks and the ruts. Although the front was jacked up as far as possible (by fitting alloy spacers between springs and turrets), we couldn't raise the tail because the axle line was *above* that of the chassis frame, and a change of frame profile was not allowed by the regulations. To minimize the damage as much as possible there was a full length undershield covering everything from the radiator to the rear of the overdrive. We also found time to add hot air vents in the bonnet sides, and (because of the puncture problems in Yugoslavia) we also mounted a second spare wheel on the boot lid, which was pressed from steel for added rigidity.

The fourth car (6 VC) was used as a high-speed 'spares on the hoof' machine, but even this car could not find John Sprinzel when his car broke down. Two of three cars survived the rough tracks as far as Novi Vinodolski, on the coast south-east of Trieste, but on the one-hour dash inland to the next control, Mike Sutcliffe's car crashed. As Roy Fidler recalls:

> I was pushing him a bit, and we came up behind another competitor in a dust cloud. Mike wanted to know if he should sit back, but I said: 'No, let's have a go, get past him.' We were going through a forest, with tall trees on both sides, and following a pair of brake lights.
>
> Then there were no brake lights, no dust—and no road! The road had turned sharp right, and we shot straight off into a field, dropped 15 or 20 feet, and a I remember seeing a Rover 3-litre in the same field, which had gone off an hour earlier; we finished up nose-to-nose, nine inches apart.

Jean-Jacques Thuner and John Gretener kept going, exhausted and dirty due to all the dust which had got into the car, to finish ninth overall. The good news was that they were the third sports car to reach the finish (the other two were Austin-Healey 3000s), but the bad news was that the car had suffered appalling damage to do this.

Not only had the undershield been badly mangled (and all the cooling slots banged

By late 1962 the 'works' TR4 interior featured special bucket seats, an overdrive switch on the facia rather than the column, the speedometer and an accurate clock ahead of the co-driver's eyes, and a revised shape of gear lever. There was also a Weber-carburated engine up front.

Jean-Jacques Thuner's TR4 tackling a forestry special stage on the RAC rally of 1962. TR4s were not ideal for this sort of motoring, but Thuner plugged on to finish 9th, and third in his class.

When my department started to prepare TR3s, we were horrified to discover that no creative homologation of TR3As had ever taken place. One of the first optional extras to be approved was a four-branch exhaust manifold, which liberated at least an extra 10 bhp.

closed), but the front chassis frame had distorted due to front suspension loads, and there had been so much load on the rear suspension that one of the leaf spring mounting pins had been impacted up through the frame, and through the floor, where the front end of the spring ended up inside the car, perilously close to John Gretener's backside.

The lessons were that we would have to treat the Liege with even greater respect in 1963, and that we should reinforce the frames even more in future. Not only that, but we asked the design office to produce sturdier rear springs (with nine leaves) for future events.

On the other hand, there was real promise on the performance front. By September Standard-Triumph had finally purchased some Weber carburettors (homologation was not necessary), mated them to SAH manifolds, and had produced around 130 bhp (and a great deal of mid-range torque) on the test bed with the 2.2-litre engine. This was less than claimed by SAH Accessories, but the factory figure was 'nett', and the results were impressive.

To try it out, we quickly rebuilt and repaired 5 VC, and sent it out to compete in Jean-Jacques Thuner's 'home' event, the Geneva rally. It was an encouraging debut, for up to the start of the very last special stage/hill-climb test he was lying second overall behind the current European Rally Champion, Hans Walter (Porsche Carrera)—when the throttle linkage parted, leaving him stranded on the line, where service crews could not reach him.

The last event of the first year was the RAC rally, which had now been transformed into a high-speed forestry stage event. The days of the 'Rally of the Tests', with navigation, which British TR specialists could have won, were over—this was another event in the fast-emerging Scandinavian pattern, where a front-wheel-drive car (and a lot more ground

A Lucas contract for lighting encouraged the team to try this lighting arrangement on the TR4s in 1962. One lamp is hidden by the tail of another car. There were two headlamps, a central 'cyclops' driving lamp, then two more driving lamps and two fog lamps. (Graham Gauld)

clearance than a TR4 ever had) was really needed.

All engines were now fitted with Weber carburettors, and the TR4 team did its best, and Thuner finished ninth overall, but if the truth be told they were outclassed, not as much by other cars as by the conditions. My lasting memory is of the number of low-mounted driving lamps which were smashed by flying stones, or by the cars nosing into water with the lights illuminated (hot glass + water = instant explosion), and of the way I kept visiting Lucas dealers all round the UK to buy replacements. Still, there was a team prize to be happy about—and there was an interesting debut by the Triumph Vitesse.

This story started earlier in the year, when co-driver Vic Elford decided that he could be a faster driver than his regular BMC driver, David Seigle-Morris, and began to prove it. By mid-summer I was receiving regular telephone calls from him, asking for a trial, so for the RAC rally Harry Webster and I decided to deal with two projects at once—to give the Vitesse an airing, and to let Vic Elford drive it.

The car chosen was 407 VC, the 1.6-litre car I had used as service transport on the Alpine and Liege-Rome-Liege events, but for the rally we entered it in modified form, in the 1.6-litre GT Class, where it would face a variety of 'works' modified Mini-Coopers and Ford Anglias.

It was an interesting gamble, for at that time Harry Webster was hoping to persuade his board to let him market a three-SU carb'd Vitesse as a cheap-and-cheerful 'homologation special' in 1963, so an early example of this 95 bhp engine was fitted. We also fitted twin 9-gallon fuel tanks in the boot—as much to hold the tail down as to increase the range—gutted the interior, and fitted $4\frac{1}{2}$ in. rim Courier van wheels (which were considered to be wide rims in those days—the standard car had $3\frac{1}{2}$ in. rims!).

For Elford, and the Vitesse, the event started well, and the handling was surprisingly good, but before half distance the gearbox (a standard Vitesse unit) began to wilt under the strain of dealing with all that power; no alternative gearbox was available—or, at least, we had not thought of using the bigger TR4 gearbox at this time.

Just before half distance, after 18 special stages of the northern loop, when the rally had trekked all the way from Blackpool, via the Yorkshire and Keilder stages, to Inverness, the Culbin stages, and back to Cumbria, Ray Henderson's mechanics commandeered a ramp at a Standard-Triumph dealer's premises in Penrith, and attempted a fast gearbox change, but it all took too long (more than 90 minutes—there were no eight minute gearbox changes in those days) and the car ran out of time. It was, at least, an encouraging way to end the season, for 1963 looked like being even tougher than 1962 had been.

Vic Elford had set so many creditable stage times—sometimes in the 'top tens'—that I hastened to sign him for 1963. As the three other drivers had all been retained, this really meant that I had one driver too many, but if I had also managed to attract Pat Moss (who was known to be leaving BMC, and eventually went to Ford), or my old colleague from Rootes, Peter Procter, a lot of hiring and firing would have been needed. Finding the money to pay them could have been a problem, but the chase was more important than the practicalities at that moment.

In the meantime Standard-Triumph Inc., of North America, had started to take motorsport seriously, and had hired a resourceful ex-racing driver, R.W. 'Kas' Kastner to prepare TR4s for customers, and for a concessionaire's team to race at Sebring. 'Kas' (no-one ever finds out what his *real* Christian names are) was born in up-state New York, moved restlessly around, following his profession as a salesman, then settled in Salt Lake City where he started racing (in an MG TD) in 1952.

Kas raced a TR2 for the first time in 1954, in Colorado, but his really important move was when he went to work for Cal Sales, the Californian distributor of Triumph cars, in 1958. In 1959 he won three racing championships, most notably the California Sports Car Club series, but suffering a heavy crash at a Santa Barbara race in 1960 he became what he described as an 'ex-race-car-driver'.

Then came the breakthrough. In 1962 Standard-Triumph's USA operation, based near

John Sprinzel brought expertise, if not outright pace, to the Triumph team in 1962 and 1963. His sleeping partner here, before the start of the 1963 Monte Carlo rally, is Sam Actman. (Graham Gauld)

New York, became more interested in motorsport. Cars such as that raced by Bob Tullius began to win their divisional races in SCCA (Sports Car Club of America) events, after which:

> Mike Cook [who was Standard-Triumph's Public Relations Manager in the USA at the time] and I were good friends. He spent a lot of time trying to get the factory interested in Sebring. Finally the importer said: 'You can go ahead, take three production cars, and build them to take to Sebring—but there's no extra budget!'

In November 1962, too, I had unwittingly laid the foundations of the ambitious race and rally programme which Triumph took up in 1964. Because the department's Vitesse 'chase' car was converted into a rally car for the RAC rally, I needed a new vehicle to follow the service cars around. The car allocated was a very early example of the Spitfire, and my report on its behaviour listed many things which would have to be changed before it could become a competition car. Less than a year later, we had started to tackle them all.

1963

For the department's second season a little more money was available. I don't recall how much, but we were still not funded in the same way as BMC and Ford.

Well before the turn of the year we had to decide what cars to enter for the Monte Carlo rally—the first to be tackled by my department. We didn't think the TR4s were right for the job—weird handicapping systems saw to that—so the only sensible alternative was to use the new six-cylinder Vitesse.

The Vitesse was at once an intriguing but infuriating design. On the evidence of the 1962 RAC rally appearance we knew we had already made it handle well, but we didn't think the engine was very good, and we knew that the transmission was weak. Not only that, but it really wasn't a good long-term prospect in the saloon car category, for we already knew that Ford was about to launch two new sporting Cortinas—the GT and the Lotus-Cortina—both of which were to be homologated soon.

At that time, naïvely, I thought that Standard-Triumph really would launch a limited-

Pre-rally scrutineering before the Monte Carlo rally of 1963, with Jean-Jacques Thuner's TR4 coming out of the door. John Sprinzel is bending down by the car, the author (in scruffy jacket and trousers) is behind it, contrasting with a very dapper Vic Elford in a smart overcoat. (Graham Gauld)

production Vitesse 'homologation special' like the car which had been used on the RAC rally, but this project was soon abandoned.

For the Monte Carlo rally, therefore, we took delivery of four powder-blue overdrive Vitesses, entered three of them, one running in the standard car, Group 1, category (for John Sprinzel to drive), the other two cars being Group 2 machines, with cylinder heads and camshafts modified free of charge by Mangoletsi, which was a tuning firm trying to ingratiate itself with the company at the time.

There was one really 'high-tech' modification on all cars. This was the fitment of an experimental Lockheed anti-lock braking system, whose sensors were operated by belts driven off a pulley on the nose of the final drive unit. Unlike modern ABS systems, the sensors only cut-off hydraulic pressure to the rear wheels when these showed any tendency to lock—the front disc brakes always operated as normal.

All the Vitesses started from Paris but, as in 1958, the weather on the Monte Carlo rally

Conditions could not have been worse for the TR4 on the Monte Carlo rally of 1963, but Jean-Jacques Thuner, who always drove 5 VC, gave it his best shot, and finished second in his class.

was awful. Only 28 of the 80 starters from the French capital reached Monte Carlo, though all three Vitesses (and Thuner's TR4) struggled through the blizzards. At the end of the event there were only eight cars unpenalized on the road sections, one of them being Vic Elford's Vitesse, though the other Triumphs were heavily penalized.

Elford's Group 2 Vitesse beat the best of the so-called Group 1 'works' Rapiers on special stage times (as an ex-Rootes team member that gave me great satisfaction), and Thuner's TR4 was third fastest overall on the Monte Carlo circuit test, but the overall results were disappointing.

After sober analysis, however, we concluded that the Vitesse couldn't possibly be a winner against Ford and Rootes unless it was given a lot more engine power. Since nothing was promised for some years to come (actually the 2-litre Vitesse didn't arrive until 1966, by which time the rally world had moved on) we decided to abandon the Vitesse project immediately.

Shortly after the rally John Sprinzel decided that he didn't want to be a team member any more, and left. A few weeks later I was asked, against my will incidentally, to lend a TR4 to Peter Bolton to enter in his local rally—the Yorkshire.

Peter Bolton was a competent racing driver, but he was no great shakes as a rally driver. Unfortunately, he had a very persuasive motor trader's tongue, which led Harry Webster to lend him the TR4. That he crashed the car only a few miles after the start of the event was bad enough—but the fact that it had been the newly reconstructed 4VC (which Mike Sutcliffe had shunted on the 1962 Liege-Sofia-Liege) only added insult to injury.

I was not amused and neither, I suspect, was Harry Webster. Ray Henderson and his mechanics were very vocal about the whole thing—and 4VC was not used again as a team car until mid-summer.

In March American-prepared TR4s made a successful racing debut at Sebring, in the 12-Hour race, though my department in Coventry was not involved.

'Kas' Kastner told me:

> I built all three cars in my two-car garage in Gardena, California. Just myself and Joe Valdes did the engines, gearboxes, differentials—everything. Then I went down to Florida thinking I was chief mechanic, but found that I was also the team manager. I was a novice, with no race team managing experience.
>
> After Sebring, where we won the class, it was a great thing for Triumph, with a lot of publici-

Mike Sutcliffe and Roy Fidler in their 1.6-litre Vitesse—powder blue with a white nose and side flashes—on the Monte Carlo rally of 1963.

Glad to be in Monte Carlo, especially as the Vitesse had struggled to be competitive in a very wintry 1963 event, are Mike Sutcliffe (rubbing his hands) and Roy Fidler (in overalls). (Roy Fidler)

ty. So I went up to New York to see the company, pointed out all the time and effort that I had put in, and suggested that someone should recognize it with dollar bills. The only reaction I got from the president was: 'Kas, if you think you want to do more, you should seek greener pastures.'

So I seeketh. Carroll Shelby talked to me, said he would love to have me work for him, and soon I agreed a deal with Ford and Shelby. He was going to pay me twice the money to go to run Shelby American with him, the Cobra plant and all their racing activities.

Even though Standard-Triumph then had a new USA president, Chris Andrews, I sent in my resignation, putting a 30-day notice on it. The only final obligation I had was to run a seminar

'Works' TR4s sometimes had less arduous duties than international rallies. Roy Fidler and John Hopwood borrowed 4 VC to use as a Clerk of the Course's 'course car' for the inaugural Manx Trophy rally of 1963. Rumour has it that Roy set fastest stage times in the process. (Roy Fidler)

for Triumph groups at Watkins Glen. When I stopped off the 'plane Mike Cook was waiting and said Chris Andrews wanted to talk to me, there and then.

I was actually in a pay booth at Idlewild Airport, New York, talking to Andrews, who said he didn't want to get into a bidding contest, but he wanted me to stay. I said: 'Well, meet the deal, I'd like to stay', and he finally said: 'OK, we'll meet the deal', and that's how I became Competitions Manager, Standard-Triumph USA. Carroll Shelby was very good about it—because I had been *that* close to joining him.

In 1963 this was an even less well-funded post than the one I was currently holding down:

For a budget we started off getting $1 a car sold in the USA—that was about $25,000 a year in 1963—out of which my salary was paid, but I had no-one else working for me. All the tools and equipment at Gardena were my own, personally owned, but I soon talked Harry Webster into giving me a dynomometer.

At the time all I could do was to persuade people to buy cars, to talk to the Zone Managers to get them the best prices for cars and parts for these guys, and I would do preparation work on cars and engines. I wrote the competition books, and then we would pay bonuses. It was a very small operation.

This, however, was the start of an eight-year relationship—in which Standard-Triumph UK learned a lot less from Kastner than they might have done. Kastner's principal skills were as a careful reader of regulations, and as a patient and successful tuner of engines within those regulations. The lasting tragedy was that by the time his achievements with the TR4 engine became known, we had already decided to drop the cars:

I never did anything for Triumph in Coventry, as such, except that I sent engines over for them to test, and stop the arguments about my figures. Do you remember the phrase 'Californian air' when it came to my claims—I had told them what power I was getting, and nobody believed me. There were good tuners in the UK getting 135 bhp on Webers, but I was getting 155 bhp on twin-SUs.

I used every word of the regulations, and worked on everything—camshaft, head, the lot. The rules I love the most are those with most words in them.

The result was that on a very limited budget Kastner soon made the TR4 *the* car to beat in its own SCCA racing classes. Bob Tullius, racing mainly in East Coast events, soon became a folk hero, and eventually SCCA national champion in the TR4's grouping. There would be a lot more of that sort of success throughout the 1960s.

Back in Europe, for the Tulip rally, three TR4s, now fitted with the third (and final) arrangement of extra driving lamps, were entered. Mike Sutcliffe couldn't do the event, so I promoted Roy Fidler to driver status in his place.

The Tulip, as ever, was a supposedly secret-route event with sprints and hill-climb tests which every works team managed to recce, and one where the class-improvement handicapping system made a mockery of the results. The TR4s were *much* faster than they had been only a year ago, and because they were now equipped with limited-slip differentials their traction out of tight corners was also improved.

Vic Elford and Roy Fidler set a series of stunningly fast times, to be third and fourth fastest *overall* on scratch times (behind the Austin-Healey 3000 of the Morleys and the Ford Falcon Sprint of Henri Greder, although Thuner went off the road twice, once actually on his 'home' hill-climb, the Faucille, near Geneva.

The handicapping system pushed the TR4s back to fourth and sixth in the GT category, but the consolation was that the three cars won the GT category team prize. Vic Elford confirmed his Monte performance by being the fastest of the three drivers—clearly he was going to be hunted by more affluent teams at the end of the year.

The team then approached the Alpine rally with high hopes, for this was an event where the latest car/driver combinations could be competitive. In the end, though, it was an

The 1963 Alpine rally was not a good event for the TR4s, all of which retired. Vic Elford, however, was fighting for the overall lead with the Austin-Healeys when he went off the road.

unmitigated disaster, in which all three cars retired, one was written off, and Roy Fidler was badly injured.

I was already accustomed to the idea of Vic Elford being (a) the team's fastest driver and (b) that he would be cocky about his prospects. I was not ready for the fact that Thuner's car cooked its clutch, Elford got so excited at the idea of beating more than one of the Austin-Healey 3000s that he put his car off the road—and Mike Sutcliffe crashed his car on the Mont Ventoux hill-climb test.

6 VC's last event with that identity—it then went on to do the Canadian Shell 4000 rally with another number—was the RAC Rally of 1963. Unhappily, Roy Fidler rolled it, and was forced to retire from the event. (Roy Fidler)

Co-driver Roy Fidler takes up the story:

We had pace notes, of course, but we hadn't been up the hill since 1962. These notes had been given to us by Vic Elford [in fact Vic and I had done a collective recce in 4 VC just before the event]. On the way up Mont Ventoux, I saw the sign for the little village, called out: 'Flat Left, then again Flat Left', and I don't recall anything after that.

They tell me it was narrow, Mike got his back wheels off the road into loose gravel, and immediately after the second bend there was a parapet with a concrete telegraph pole. I then woke up in hospital.

Fortunately for Roy, his broken left leg mended very quickly indeed, though he lost a small toe. He was back in rallying in the autumn, and spent some time driving an automatic transmission car as soon as he returned home from a French hospital.

By this time, I'm sure, Harry Webster and I knew that the TR4 was coming towards the end of its 'works' career. Although we knew that the mid-1963 cars were a lot better than the early-1962 examples, we also knew that there was nowhere else to go. The engineers at Coventry couldn't get any more power from the engines—'Kas' Kastner had not then shown how he could get a lot more power with 'Californian air'—which meant that we could not match the Austin-Healeys, Porsches or Alfa Romeo TZs.

We had made up for all the time lost by the previous department, but since motorsport never stands still it was all a bit too late. There were many stages and hill-climbs in Europe which never changed—same starting lines, same finishing lines, same tarmac—and I eventually compared 1963 TR4 times with those of other cars in previous years.

The conclusions were startling—if 1963 TR performance had been used in 1960 and 1961 (which was perfectly sporting-legal, even for the TR3As), the cars would have been faster than the Austin-Healey 3000s of the day, which 'only' had about 180 bhp. It was nice to know that the TR4 was good—but it was no longer a winner by 1963 standards.

It was at this point that strategies for 1964 began to be discussed—and these are detailed in the next chapter. For the next few months, however, we had to tackle two more events. It wasn't an encouraging prospect, for both were rough, loose-surface, rallies which would probably batter the TR4s to pieces. That was the prospect, and it was borne out by the facts.

Here's a rare and interesting shot—Brian Culcheth and Roger Clark (yes—that Roger Clark) in a 'works' TR4 at the start of the 1963 Spa–Sofia–Liege event. It was the only occasion Roger drove a 'works' TR4, though he would eventually campaign a TR8 in 1980. (Brian Culcheth)

For the 'Liege', which had achieved its final title of Spa-Sofia-Liege, we entered four cars—three TR4s as usual and a Vitesse prototype for Vic Elford, for the regulations allowed special cars to be used. There were several crew changes. I hope Mike Sutcliffe understood why I dropped him after the Alpine rally—it was not the only accident he had had with the team, and we knew his eyesight was not what it should be—and because Roy Fidler was still recovering from injury, we had to find two new crews.

My agreed strategy was to use Roy again as soon as he was fit (his Tulip rally performance had been staggering), and since he was planning to take Don Grimshaw as a co-driver, I allocated 3 VC to him; Don had recently been very successful as a driver in British rallies, in a TR3A, and later in an ex-works Austin-Healey 3000. For the vacant car, I took a big gamble, offering the newly-rebuilt 4 VC to Roger Clark and Brian Culcheth, both of whom were young British drivers at the time.

Both Culcheth and Clark were just breaking into international rallies, and both had performed well in the recent Alpine rally, where 'Culch' had won his class in a Mini-Cooper, and Roger had been second in his class in a Reliant Sabre Six. Not only that, but Roger had been outstanding in recent British events in a Mini-Cooper. In my own small way, I was happy to entice these people out of a BMC car and into a Triumph. BMC's Stuart Turner, who was already a good friend of mine, cheerfully pointed out that he had a surplus of world-class drivers anyway and didn't mind at all!

Brian Culcheth said:

> Roger and I were good friends, and we were both wanting to break into works teams. I'm sorry to say that we didn't get very far on this rally though, as the car broke down on the Pec-Titograd section.

The story of the Vitesse which Vic Elford took on the Liege really deserves a chapter to itself, so I must compress everything a lot. Vic wasn't happy at the idea of pitching a TR4 against the rocks of Yugoslavia (he had done the same as David Seigle-Morris's co-driver, in an Austin-Healey 3000, in 1960, and hadn't enjoyed it one bit), so started lobbying for a modified Vitesse instead.

Earlier in the year he had used a triple-carburettor car (6003 VC) on the Manx rally, and had been leading the event until the differential stripped its teeth. That was a surprise to no-one—it merely confirmed what the RAC rally outing had already suggested.

At that time of the season, money was short. When Vic asked if we could build an even-

Vic Elford (left) and David Stone spent a season with Triumph in 1963, their first season-long contract, which led to even greater things in the future.

more modified Vitesse for the Liege, I had to refuse him. He then approached Harry Webster, who also refused him. Nothing daunted, Vic then approached Standard-Triumph's new chairman, Donald Stokes, and asked *him*. I watched these manoeuvres with polite interest, knowing that we didn't have the money to do the job, but that if Vic *did* get his way, then the money would suddenly appear.

It did. The silver-tongued Elford (he was a life assurance salesman at the time!) persuaded Stokes that he, Vic Elford, could win the event, and all of a sudden we were cleared to do the job.

To make the transmission bomb-proof we effectively started again with 6003 VC, for we installed a 2-litre/110 bhp engine, a TR4 clutch, gearbox and overdrive, and a new-style TR4A/Triumph 2000 chassis-mounted differential, along with all the related components we thought were needed, including a larger radiator, a beefed up drive line, and a new-style front end with single 7 in. headlamps instead of the twin 5 3/4 in. lamps normally carried.

There was no chance to test the car—it was not actually finished until the mechanics were on the cross-channel boat, on their way to the start of the rally itself—although we thought the handling and cooling would be right for the job.

But there was no happy ending. Half way through the rough race back from Sofia (in Bulgaria) to Italy, by way of Titograd and the Adriatic coast, Vic Elford and his new co-driver Terry Hunter were in third or fourth position, but near Bribir the car suddenly caught fire (it was *not* crashed) and was completely burnt out, happily without injuring the drivers. The wreckage was total, so we can only guess that there had been some sort of rupture in the fuel lines, or that a carburettor had sprung a leak.

This year, too, the TR4s all retired, battered into submission by the horrible conditions. Brian Culcheth told me that:

> When the TR4 broke down on the Pec-Titograd section, we were utterly and totally exhausted. We both sat down by the car and fell asleep, and woke up some dusty hours later. We got the car back to the nearest village—we got towed down by the local bus—and as soon as we found the hotel we discovered a tremendous party going on, with Timo Makinen, Geoff Mabbs and others, who had already retired, enjoying themselves.
>
> We'd broken the gearbox, so to get the car back to England we put it on a train, and went home on another train.

In the meantime, 'Kas' Kastner had sent over the first of his famous 'shopping lists'—asking

Three 'works-USA' TR4s lined up at the start of the Sebring 12 Hours race of 1963, in which they dominated their class. Note the ultra-wide disc wheels. ('Kas' Kastner)

Above *Bob Tullius (at the wheel of this TR4) made his name as a racing driver in the early 1960s, and was Triumph-USA's only 'works' driver for some years. Even in 1962 he was carrying his habitual competition number—44.*

Above right *'Kas' Kastner was Triumph-USA's competitions manager throughout the 1960s, and a man who produced miracles from limited resources. Here he is posing in front of one of the three victorious TR4s at Sebring in 1963. ('Kas' Kastner)*

for certain motorsport options to be nominally listed for the 1964 USA-specification models, so that he could upgrade the race cars, and make sure that they continued to dominate the SCCA D-Production category.

'Kas' was also invited to visit Standard-Triumph in the UK, and remembers how generously he was received by Harry Webster in Engineering ('I thought Harry was, and is, terrif-

Bob Tullius (TR4), not only carrying the competition number 44, but also carrying Quaker State sponsorship, was a very important part of the Triumph-USA scene in the early and mid 1960s.

ic'), but how he was almost ignored by the sales staff when invited to visit them at the Earls Court motor show.

Kastner said:

> We struggled at first for we had to fight against the BMC cars, which already had a ton of options. But we were always competitive—and eventually we ran over them like a tank running over a rabbit.

Autocar borrowed 6 VC for a road impressions test after the Tulip rally of 1963, when it was in fully-stripped out condition (except for the large-capacity 18-gallon homologated fuel tank), and was running with the 4.3:1 'Alpine rally' rear axle. It was interesting to note that the rally car, when equipped with its tool kit and all the usual fittings, was still heavier than a standard car—what it would have been without aluminium panels, heaven knows.

Except that writer Stuart Bladon credited the car with having glass-fibre front wings (these were, in fact, in aluminium, as usual on all the TR4s), it was an accurate and carefully detailed test.

We had never actually found time to 'figure' the car against the stopwatch, so *Autocar*'s figures, which showed 0-100 mph cut from 46.3 sec. (standard car) to 26.4 sec. (4 VC) were illuminating. Acceleration in the upper rpm region was startlingly better than standard—40 mph to 80 mph in direct top occupied 12.9 sec. instead of 19.1 sec.—and showed why the fully-developed TR4 rally car was so effective on tarmac. If only international rallying had not changed so much in the early 1960s.

As it happened, for Standard-Triumph the 1963 season ended with a whimper. Unlike 1962 we couldn't even get three TR4s back to the finish of the RAC rally, and have a chance of winning the team prize, for Vic Elford's car blew a cylinder head gasket (that was almost unheard of in a wet-liner TR4 engine, so we put it down to Murphy's Law), while Roy Fidler, who had made a remarkable recovery from his broken leg in June, rolled his car:

> It was on the way back through the Lake District towards the half-way halt. I was tired, and I remember taking a Proplus tablet. That made me feel very confident, I thought I could do anything with the car—then I lost the back end on a left-hander, it put a wheel over the edge, then went sideways, then rolled over and slid along on its roof. I was disgusted with myself.

By the end of 1963, therefore, Triumph faced big changes in the motorsport department. The TR4s were about to be pensioned off, our young star driver, Vic Elford, had been attracted away to drive for Ford—and new cars, and new programmes were all in the wind.

Chapter 5

Big budgets and high ambitions
1964 and 1965

By the autumn of 1963, Harry Webster and I were both getting very frustrated. Although the TR4 had been turned into a fast and reliable car—it was already a much more capable machine than the TR3A had ever been—it didn't look as if it could be turned into an outright winner. In 1963, on events with gravel stages it was not competitive with the latest breed of front-wheel-drive cars. On tarmac events where its grip and traction were excellent, it was usually handicapped out of the running by peculiarities in the regulations.

It was almost as if the fates were against the TR4. Even if the Kastner magic (and what 'Kas' whimsically quipped as 'Californian air'!) could be breathed into engines without infringing homologation rules, the TR4 would probably not be able to beat the Austin-Healey 3000s in any case. Worse—Porsche had just launched the rear-engined 911 which, if rallied, would be in the same class and category as the TR4.

It was time to think again, and for the second time in less than three years Harry Webster invited me to look into Triumph's motorsport future. He needed no convincing that his cars should be in 'works' motorsport, but like the good businessman that he was, he wanted to see the alternatives spelt out.

Compared with 1961, however, the background to our discussions was much more favourable. In 1961 Standard-Triumph was recovering from a traumatic takeover, but in 1963 its self-confidence had returned. In 1961 the company was losing money hand over fist, but in 1963 it was profitable once again. In 1961 the competitions department had been closed down, but in 1963 it was active.

Harry asked me to summarize my thoughts on paper, promised to study these very carefully, talk them through with me, then decide what he should recommend to the board of directors.

The problem was that, in the two short years of rallying with the TR4s, motorsport in general, and rallying in particular, had moved on rapidly. BMC, Ford, Alfa Romeo and Renault had all produced early examples of what became known as 'homologation specials'; Standard-Triumph, a smaller company with fewer resources and a crowded forward-model programme, had no intention of matching them.

Straightaway, therefore, we had to ask ourselves whether we could be competitive when battling against low-production 'specials', or whether there were ways of getting round the 'homologation special' dilemma.

Further discussion with Harry Webster helped to identify three courses of action for 1964 and beyond. These were:

    (a) To carry on rallying TR4s which, even with more development work, were not likely to be competitive.

    (b) To close down the department, citing the 'homologation specials' and the need for heavy spending to match them, as the reason.

(c) To expand considerably, and to commit more money and resources to the development of
     new models.

Every Triumph enthusiast should thank Webster for rejecting the first two options. Harry,
the great enthusiast, wanted to take Triumph back to Le Mans, and to see a bigger programme in 1964 and beyond.

For my part I emphasized the way that more and more rallies were taking to unmade
tracks, some of them very rough, and that handicapping which favoured smaller-engined
cars seemed to be well-established. No matter how the sport changed and developed in the
next two or three seasons, it seemed to me that no single Triumph model could be right for
every occasion.

There was a lot of discussion. We analysed what was already available—Triumph had the
Herald, Vitesse 1600, Spitfire and TR4, while the Standard marque had just been laid to
rest—and we knew that the big 2000 saloon would go on sale at the beginning of 1964. In
1965 there would be a new TR4A, which had independent rear suspension, and the prototype 'Spitfire GT' (complete with fastback coupé body style) had just been completed.

There wasn't going to be a problem in getting homologation when we needed it, as all
these cars were, or would be, in quantity production.

In addition there were obvious outside influences. Healey of Warwick, in conjunction
with BMC was already doing marvellous things with Austin-Healey Sprites and MG
Midgets. Mercedes-Benz, Ford, Rover and—to a lesser extent—Humber had all shown that
a large saloon car could be spectacular and successful in loose-surface rallies.

Looking back from the 1990s, Triumph's motorsport department experienced two miracles at about the same time in autumn 1963. One was that Harry Webster agreed to recommend a two-pronged rallying attack, and the other was that he decided to recommend a
return to the Le Mans 24 Hours race.

He agreed to recommend that not one, but three entirely new sets of cars should be prepared for use in 1964, and that a great deal more effort (and money for development and
operation) should be put behind them. Two sets of cars should be Spitfires (or radical evo-
lutions of that theme), while the other set should be of Triumph 2000 saloons.

*This was certainly the first 'works' Triumph 2000—a pre-production car driven by Colour
Sergeant Rhodes in the London Motor Club's TV Autopoint in the winter of 1963/1964. Apart
from some undershielding, it was almost entirely standard. (Gordon Birtwistle)*

The intention was that the Spitfires would be confined to tarmac rallies, and race tracks, while the 2000s would be turned into rough-road 'tanks' for events like Spa-Sofia-Liege, or the RAC rallies. This approach faltered in their second season (1965) when the 2000s were more widely used on events like the Tulip and Alpine rallies, though the 'works' Spitfires were never obliged to drag their bottoms over un-made tracks.

Clearly there was general agreement at management level that this sort of programme was justified on marketing grounds. Backed up by George Turnbull, and by John Carpenter in the sales department, Harry Webster gained board approval for this programme in December 1963—though it was typical of Webster that he had already approved the start of design work, and of informal approaches to various factories for special components to be produced, several weeks earlier than this!

At the time I never had to deal with budgets, or the monitoring of them (today's competitions managers must envy me that, at least!), so I cannot quote exact figures for the cost of the 1964 programme. In any case, as in previous years quite a lot of work could be 'hidden' in other budgets—if engine development work for the racing Spitfires was being done, this could be 'lost' in the general development budget for Herald and Spitfire engines, for instance. However, for 1964, there may have been three times as many cars, but there was at least eight times as much money!

Because we knew we could not compete on equal terms with the new breed of 'homologation specials', especially in the saloon car categories, we decided to avoid them wherever possible. Even in hindsight, I like to think that we were quite thoughtful in all this.

For the Le Mans 24 Hours race, the Spitfires were to run as 'Prototypes', and not as homologated cars. In their first year, therefore, this would give the engineers freedom to make the cars as rugged and reliable as possible.

For rallying, the Spitfires would be modified to the very limit of the regulations, not only by lightening them, but by changing their shapes, and by using what we would now call 'inventive homologation' to get the best possible power outputs, and the most suitable transmission options.

The big 2000 saloons, though sturdy, were likely to be underpowered in standard form, so it was decided to run these as what was known as Group 3 (or Grand Touring) models, which meant that although they would have to run against GT cars like the Porsches and the Austin-Healeys, they could have highly tuned engines, non-standard wheels, brakes and suspensions, and—as with the Le Mans Spitfires—they could be idealized for their jobs.

Not only that, but it was decided to split the location of the 'works' cars into two areas. The rally cars were to be prepared in the same department as the TR4s—in a workshop at the eastern end of the Fletch. North engineering block—while the Spitfire Le Mans race cars would take shape in the heart of the experimental department, close to John Lloyd's office. Further, I would look after the rally programme, reporting direct to Harry Webster, while John Lloyd (who was in any case Harry's deputy) would look after the race car project all on his own.

Compared with the close-down of 1961, and the cash-strapped seasons which had followed, this was a heartening improvement in commitment. All we had to do now was to deliver the goods.

1964

As expected, there was a change to the team of drivers. We had lost two drivers—one of them in the 'driver transfer market', the other due to my neglect. Vic Elford had defected to Ford, while I had not been able to give another drive to Roger Clark in the autumn of 1963. By the time I knew that Vic would be leaving, Roger had been snapped up by Rover—and the rest, as they say, is history.

Which left us with only two drivers—Roy Fidler and Jean-Jacques Thuner, but I soon rectified this by retaining Rob Slotemaker for Le Mans and the Tour de France, and by signing Terry Hunter (Vic Elford's 1963 Liege co-driver).

Happily, the various event dates could not have been spaced at a better time. There was a lull early in the year when cars could be built, when testing could begin, and when dramas and problems could be uncovered. The TR4s would make their last 'works' appearance—in Canada—in April, the Spitfire rally cars would race at Le Mans in June, the Spitfire rally cars would appear on the Alpine rally a week later, but the 2000s were not scheduled to make their first appearance until the Spa-Sofia-Liege started at the end of August.

Although Terry Hunter tried to persuade me to lend him a car to compete in the Monte Carlo and Tulip rallies, there was no way that this could be done. Early in 1964, therefore, Ray Henderson's mechanics had to concentrate on re-preparing three of the four TR4s to leave for Canada, to compete in the Shell 4000 rally, and Spitfire preparation could not begin until that had been done. The TR4s had taken a battering on the RAC rally of November, so even more work than usual was needed on the cars.

Three cars—previously registered 3 VC, 5 VC and 6 VC—had those identities removed and were completely rebuilt, around new chassis frames. In all cases they were converted from right-hand to left-hand drive and they were fitted with standard disc wheels, though their basic specifications—2.2-litre engines with twin dual-choke carburettors, rear axles with limited-slip differentials, full-length undershielding, aluminium body skin panels, per-

'Kas' Kastner of Triumph-USA making final adjustments to one of the three TR4s sent out from the UK to compete in the Canadian Shell 4000 rally in 1964. This is one of the few pictures I have ever seen which shows the installation of two dual-choke Weber carburettors on the 'works' TR4s, the use of a brake servo, and the air horns on the left inner wheel arch.

This shot is a real rarity! The three 'works' TR4s lined up in the Triumph-USA competitions department in April 1964. Before being rebuilt as left-hand-drive cars, these had been the UK-based cars 3 VC, 5 VC and 6 VC. They were about to compete in the Shell 4000 rally of Canada. ('Kas' Kastner)

spex side and rear windows—were all retained. Memories fade, but I have recently been told by historic TR enthusiasts that the cars were also given new chassis numbers appropriate to their specification and their next destination—the USA.

When completed in February, the cars were driven to Immingham, where they were

During the Canadian Shell 4000 rally, the author (in sports jacket) and Triumph-USA's competitions manager 'Kas' Kastner followed the TR4s all the way across the continent in a Chevvy van, full of tyres, wheels and parts.

The last appearance by a works TR4 was in the Shell 4000 rally of April 1964. The cars were run without bumpers (the regulations allowed them to be removed), had special USA-built wheels, while the plastic rear windows carried all the stickers from previous events tackled in previous seasons. ('Kas' Kastner)

loaded on to a freight ship which was also taking thousands of Spitfires, TR4s, MG Midgets, MGBs and Austin-Healey 3000s to North America. In the case of the 'works' cars, they were then unloaded at Long Beach, after which they would be delivered to 'Kas' Kastner's workshops in Gardena, Los Angeles, for final preparation.

In California, not only did 'Kas' run an expert eye over the cars (which were to be entered in the Canadian Shell 4000 event by Standard-Triumph Inc. of the USA) but he had them registered, and had them fitted with cast alloy road wheels in place of the standard disc wheels on which they had been transported.

Neither 'Kas' nor I can now recall why these cars were registered CAG 408, CAG 409 and CAG 410 in Oregon. Gardena, after all, is in California, S-T Inc. was based on New York State, while the nearest state to Vancouver in Canada, where the rally was due to start, was Washington!

The Shell 4000 entry had been made with publicity in mind, which explains why there were journalists in the team, and no-one who was an expert in precise time/distance rallying, North American style. The fact that there were some speed sections along the route, which started in Vancouver and finished in Montreal, meant little.

Well in advance of the event I warned Harry Webster and Chris Andrews that we could hope for nothing—and so it transpired. Although Jean-Jacques Thuner was fastest on some of those stages, none of the crews came to terms with the navigation, which explains why conventional cars such as a Volvo 122S, a Chevrolet Chevvy II and a Ford Falcon dominated the results. All in all, it was a miracle that three cars finished, and won a team prize; Roy Fidler, co-driving for Jean-Jacques on this occasion, described the computations needed as 'a nightmare'.

This really was the end of the UK-built TR4 story, for those three cars were then sold off in North America, while the fourth car (4 VC) was bought by Gordon Birtwistle. 4 VC was later bought by an enthusiastic member of the Triumph TR Register, while one of the ex-Shell 4000 cars was advertised for sale, from the USA but in British magazines, in 1991.

Well before this, Roy Fidler had competed in several British events in the Spitfire development car (412 VC) with the rather depressing result that he retired on most of them, usually with a blown engine. This, in fact, was the principal problem with Spitfire development

412 VC was a very early Spitfire prototype, built in 1962, but in 1964 it was turned into a very useful rally test car, which allowed parts for the blue 'works' Spitfires to be developed. Roy Fidler and John Hopwood are on the start line of a special stage in the Welsh International rally, in which they finished second overall.

at the time—for it was clear that about 100 bhp was needed for the cars to be competitive, and it was very difficult to produce reliable engines at first. The first-ever endurance test at Oulton Park had to be abandoned because of engine failure.

In the meantime, no fewer than nine new 'works' Spitfires—four race cars and five rally cars—were being built, and made ready for the Le Mans race and the Alpine rally respec-

The first test of a Spitfire Le Mans car was carried out at Oulton Park in the spring of 1964, with Triumph's test driver Fred Nicklin, plus David Hobbs and Peter Bolton at the wheel. In this debriefing session (left to right) are Fred Nicklin, Peter Bolton, David Hobbs, and the author.

tively. Although the chassis and engine technology, and the homologation work put in, was the same for both types of car, there were several basic differences in 1964.

Race cars were immediately equipped with all-aluminium shells to which glassfibre fast-back tops had been grafted, while rally cars had basic steel shells, aluminium skin panels, and normal bubble-top hardtops. Race cars used TR4 gearboxes while rally cars used close-ratio Vitesse type gearboxes. Race cars used alloy wheels, while rally cars used wide-rim steel wheels. Race cars used cast iron cylinder heads, while rally cars used aluminium heads.

The Nicklin/Birtwistle partnership eventually produced good handling cars.

Gordon told me:

> We made the Le Mans Spitfires handle by making the suspension very stiff, so that they hardly rolled at all. But it only postponed the inevitable 'jacking-up' problem to a much higher corner-ing velocity. I recall that when Robbie Slotemaker drove at Le Mans, he got black-flagged! He used to get quite sideways in the Esses, and when Olivier Gendebien couldn't get past in a Ferrari he complained. The stewards didn't like his driving style.

Two race cars were made ready for the test weekend at Le Mans in April, where they proved that the latest engine modifications had produced a reliable unit. At this stage the 'E-Type' faired-headlamp bonnets were not available, yet the cars lapped in 4 min. 54.7 sec. and 4 min. 54.8 sec., which was a mere eight seconds slower than the 1961-spec. TRS cars. In race trim in June, with faired-lamp bonnets and with partly blanked-off front grilles, they weighed about 1,630 lb., and reached 134 mph, at which point the engines were singing along at a steady 7,000 rpm. This was remarkable, for the Spitfires were only 1.1-litre machines with 100 bhp—which proved that preparing smaller cars, with good aerodynam-ics, had been a good gamble.

All four new race cars were taken to Le Mans in June, and three of them started the race. This effort was controlled by John Lloyd and Lyndon Mills, with Harry Webster in atten-dance. Not only did they have much more experience of long-distance racing than anyone

Preparation of Spitfire rally cars about to begin in the spring of 1964. The four bodies are of the powder blue cars, but there are five chassis on trestles—the fifth being readied for use under Valerie Pirie's SMART car.

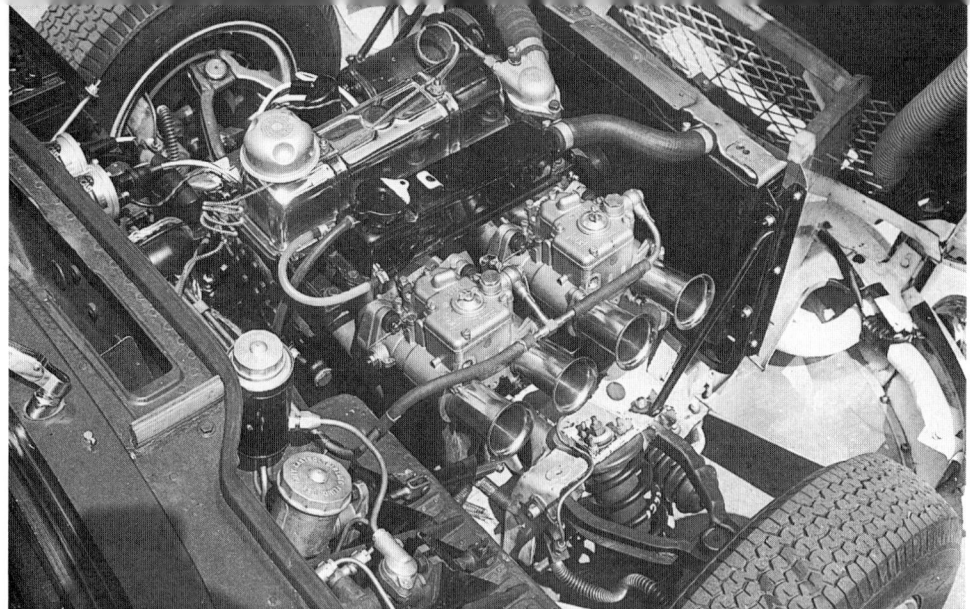

The Le Mans Spitfire's engine bay was certainly full of machinery. This is an April 1964 study, actually taken at the Le Mans circuit. Other obvious details are the armoured brake hoses to the front disc brakes, the twin ignition coils mounted on the bulkhead and the separate brake master cylinder reservoir.

else at Standard-Triumph, but there was in any case a near-clash with the Alpine rally, where I had to control the rally team.

To the surprise of almost everyone except the engineering and motorsport team which had developed the cars, the Spitfires themselves were almost totally reliable at Le Mans—although the drivers were not. After only two hours Mike Rothschild crashed ADU 1B under the Dunlop Bridge (possibly because it had been pushed off-line by a faster car), while after more than 12 hours there was a more serious accident when a woozy Jean-Louis Marnat crashed ADU 3B just beyond the pits.

In April 1964 the Spitfires went to the Le Mans test day, a weekend marred only by the awful weather, and by the tendency for the painted competition numbers to come adrift! At this point, too, the faired-headlamp noses had not yet been completed.

One of the aluminium-bodied racing Spitfires in the pits at Le Mans, during the test weekend of April 1964. There was still a lot to be done before the cars were considered race-ready.

Gordon Birtwistle, who was working in the pits, remembers what happened at 04.30 in the morning:

> He had been hit up the back by another car [at Tertre Rouge], and the rear panel which gave access to the spare wheel had fallen out. That's why he was getting exhaust fumes into the car, which were affecting him, but he didn't know that.
> The car came past us, weaving down the straight, not going very fast, and it just glanced off the end of the pit counter close to the Alpine-Renault team, then went over and buried itself into the foot of the Dunlop Bridge on the outside of the corner.

The third car, driven by David Hobbs and Rob Slotemaker, circulated speedily and totally reliably for 24 hours, averaging 94.7 mph, and finishing 21st overall. Triumph were sorry to

The pre-race line-up at Le Mans in 1964 featured three Spitfires, all proudly carrying consecutive registration numbers. Only one of these cars—car No. 50 (ADU 2B) would finish the race.

The Spitfire rally cars only appeared once with 'bubble-top' hardtops—in the Alpine rally of 1964. ADU 7B was usually driven by Terry Hunter in 1964, with Patrick Lier as his co-driver.

have been beaten by two 'works' Alpine-Renaults, but on the other hand the team was delighted to have thrashed the 'works' Austin-Healey Sprite by a full 124 miles. It was an encouraging start to a programme which had only existed on paper in January, and for which the first cars were only finished in April.

In the meantime, the rally team was preparing for its first event with the Spitfires—the Alpine rally—but with an added burden. Early in the year an extra Spitfire programme had been imposed on the small department. Stirling Moss had a racing team (SMART) and he also had an ambitious secretary, Valerie Pirie, who wanted to become a rally driver.

Moss's oil company sponsor was BP, Standard-Triumph's oil company sponsor was BP, and before long a complex deal was worked out whereby my department would build and maintain a car for her (ADU 467B), she would enter the same events as the 'works' team, but BP (and Standard-Triumph Public Relations) would pay for the car to be run.

This was to be a 12-month deal, starting with the Alpine rally, but there was never any question of Val gaining a place in a 'works' car at the expense of a regular driver. The result was that a brand new fifth car was built, and painted in that rather awful green colour which Moss's favoured (I once called it 'Channel passenger green', and had my knuckles rapped for being flippant.) The big difference, though, is that this car was never rebodied with the fastback style of the official cars; throughout the 12-month period it ran with a normal removable steel hardtop.

At the time, this deal would only pay dividends if the car had been successful—but it was not. Looking back, it seems to have been a complete waste of BP's (and Stirling's) money, for although Valerie was a good influence on the *real* works team (she was always cheerful and full of fun), she was not in the same class as a driver. She crashed on the Alpine, blew the engine on the first race in the Tour de France, went off the road on the RAC rally, fell back and ran out of time in the 1965 Monte, and finished, well down, in the 1965 Tulip. Five events, one finish, and no awards—was this what the public relations experts had expected?

It was on the way down to the start of the Alpine that we realized that these engines would always be pigs to start from cold, this being due to the use of a tiny 10 mm type of racing motorcycle plug, a very high compression ratio, and rather a puny starter. There *was* a successful technique, which we eventually learned, but this didn't evolve until we'd seen

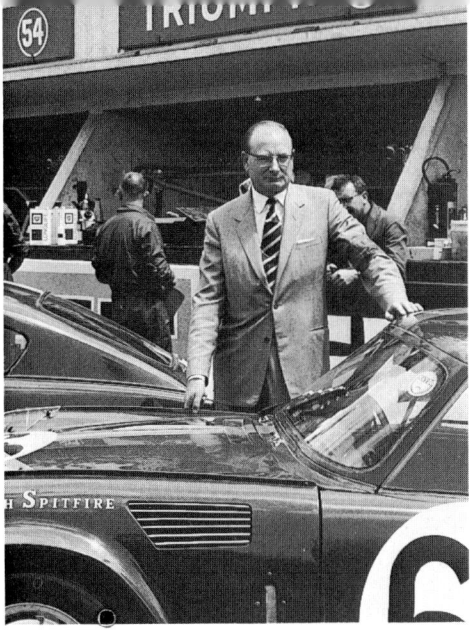

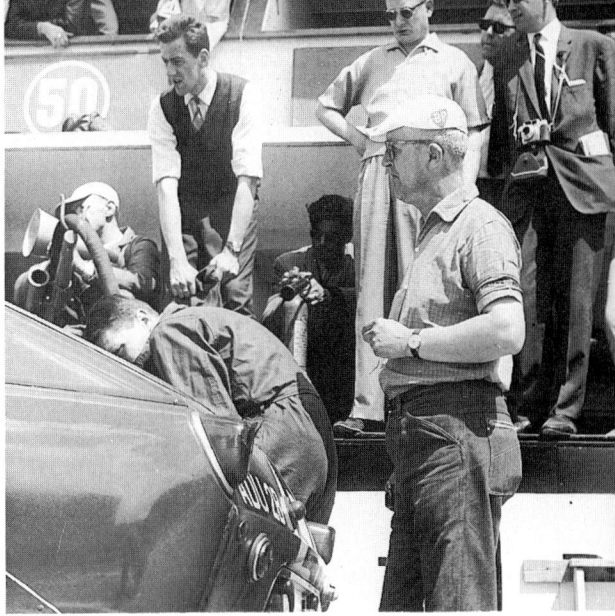

Above *Harry Webster standing proudly behind one of the three British Racing Green Spitfires at Le Mans in 1965.*

Above right *Triumph personalities in the pits at Le Mans in 1964 include Fred Nicklin (in overalls), Ray Henderson (with megaphone), Gordon Birtwistle (in dark pullover), Harry Webster (in dark glasses and short sleeves), George Turnbull (behind Webster's left shoulder), and Pierre Blanc (camera round neck—he was Triumph's Swiss importer). (Gordon Birtwistle)*

the cars being pushed off the Motorail flats at Lyons because they could not be started after a night in the open air!

As at Le Mans, the 'works' team only got one of its Spitfires to the finish, but there was a great deal more variety in the reasons for retirement. Jean-Jacques Thuner's car retired after a head-on shunt with a Belgian tourist on the Col de la Cayolle, while Roy Fidler's car damaged a piston and immediately threw a con-rod through the side of the cylinder block. After the event:

Below *The Marnat/Piot Spitfire was eliminated at Le Mans in 1964 when it crashed, after an earlier accident had broken rear body seals and J-L. Marnat passed out due to carbon monoxide poisoning.*

Below right *Change-over time for drivers during a pit-stop at Le Mans in 1964. David Hobbs about to get into a fly-bespattered Spitfire.*

Roy Fidler and Don Grimshaw about to start the Alpine rally from the seafront in Marseilles in 1964. That is an engine oil cooler mounted behind the front grille. (Roy Fidler)

Roy reminded me:

Harry Webster called me down to see him. I went into a room with a huge table, and I thought: 'This is it. I'm going to be sacked for over-revving the engine.' Harry then pushed a broken con-rod across the table and said: 'Look at that—I want to thank you for this. See—there was a weakness there, it wasn't your fault, there was a weakness in these rods, and we're changing the production line because of this. I thought you'd like to know.'

Terry Hunter should have won a *Coupe des Alpes* on this event, and I blame myself for his loss of a single minute which took it away from him. We committed two sins—one was to hold him for too long at the service point at Entreveux, the second was to under-estimate the time it would take him to reach the next control on a section where we knew the time allowance had been 'pruned'; only seven cars actually won *Coupes* that year, one of them being the Spitfire's class winner, Rauno Aaltonen.

Terry Hunter and Patrick Lier competing in the Alpine rally of 1964, in which they missed an Alpine Coupe by just one minute.

The best result ever achieved by the little Standard 8 was in the Tulip rally of 1956, when team cars finished second, third and fourth overall. Here is the 'bearded wonder' — Johnny Wallwork — bustling on his way to second overall. (David Weguelin)

The Works Triumphs in colour

This very rare shot shows a 1959 Triumph TR3S going through pre-event scrutineering at Le Mans. It was driven in the event by Peter Bolton and Mike Rothschild. (Keith Ballisat)

Above *Simo Lampinen's left-hand-drive Triumph Spitfire in 1965, with Simo at the wheel. Running as No. 3, with a 1.3-litre engine, this car won the entire prototype category in the Alpine rally of that year. (Gordon Birtwistle)*

Facing page *R.W. 'Kas' Kastner was the Triumph-USA competitions manager for most of the 1960s. He later went on to run the very successful Nissan-USA IMSA racing team in the 1980s. ('Kas' Kastner)*

Below *John Sprinzel at the wheel of the Triumph 2000 during the Alpine rally of 1965. His co-driver Willy Cave is outside the car, talking to Ray Henderson. (Gordon Birtwistle)*

'Kas' Kastner's K-Car, styled by Peter Brock, was raced once at Sebring in 1968, then abandoned.

The entire British Leyland team before the start of the 1970 World Cup rally — four Triumph 2.5PIs, two Austin Maxis and a Mini 1275GT. The Triumph crews, clockwise from the lower left are Brian Culcheth/Johnstone Syer, Paddy Hopkirk, Neville Johnstone and Tony Nash, Lacco Ossio, Andrew Cowan and Brian Coyle, and Hamish Cardno, Evan Green and 'Gelignite' Jack Murray. Culcheth's car finished second, and Hopkirk's car fourth. (Alan Zafer)

The first overseas appearance for the fast-developing Dolomite Sprint was in Belgium — the 24hr of Ypres rally of June 1974. The crew are Brian Culcheth and Johnstone Syer, the car is the hard-working FRW 812L. (Hugh Bishop)

By 1977 RDU 983M was nearly three years old, but every part of the original car had been replaced by that time. This was its last-ever outing, where Pat Ryan drove it to 12th overall on the Scottish rally, and shared the Manufacturers' Team Prize with the two 'works' TR7s. (Hugh Bishop)

All the glamour of a European rally in the late 1970s, with Tony Pond and Fred Gallagher ready to start the Elba Rally in 1977. They finished third overall. (Hugh Bishop)

A famous occasion! This was the TR7 V8's first international victory, when Tony Pond and Fred Gallagher won the all-tarmac 24hr of Ypres rally in Belgium in 1978. (Hugh Bishop)

When the TR7 V8s were on song, they could be spectacular and very effective, especially on tarmac. This was Eklund's car on the San Remo rally of 1979.

In the 1980s one of the surviving Dolomite Straight Eights was beautifully restored, in the form in which it competed in the Monte Carlo rallies of 1935 and 1936. (Mirco Decet)

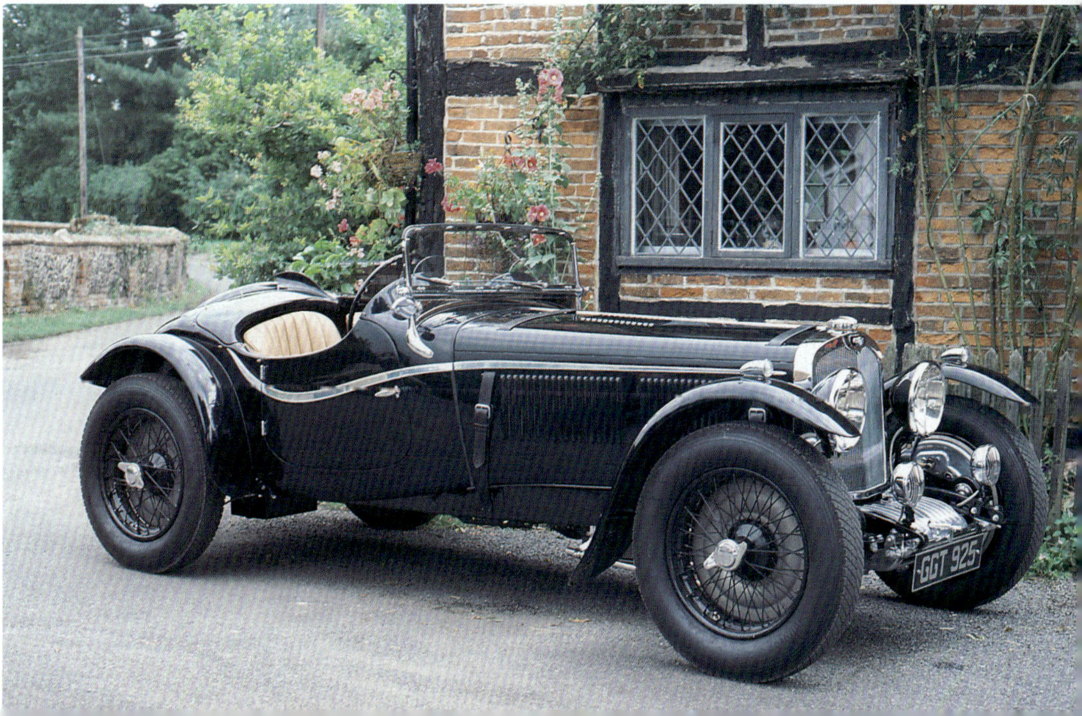

The 'works' TR7 V8's last major success was an outright victory in the Manx International rally of 1980. Tony Pond and Fred Gallagher were ruthlessly competitive in these cars on tarmac events, where a combination of 300 bhp and excellent traction made them near unbeatable. (Hugh Bishop)

Fifty years on, the restored Dolomite Straight Eight on the move, with Tony Merrick (whose business produced this remarkable renovation job) at the wheel. The wheels really were as large as they look. (Mirco Decet)

Then it was time for a chance of emphasis—from super-tuned Spitfires to big and hope-fully solid five-seater 2000s. Although we were all looking forward to developing a team of Triumph 2000s, we had virtually no experience to draw on when work began in the spring. Except for the one-off use of Standard Vanguards in the 1956 Monte Carlo rally, Standard-Triumph had never entered its large saloon cars in motorsport.

In motorsporting terms, no-one knew anything about the Triumph 2000s; the only proven competition item in its chassis was the gearbox/overdrive, which was the same as that used in the TR4. Except that Colour Sergeant Rhodes had driven a pre-production car on the London Motor Club's TV Autopoint event in December 1963 (by that time it was traditional that Triumph lent a vehicle to the BAMA for this occasion), the cars had never been driven in anger.

On the other hand, we could at least draw on the results of all the rough-road and *pavé* development testing done on prototype cars in 1962 and 1963, and we had already started using heavily overloaded 2000s as service cars earlier in the season.

This was where Harry Webster's engineers, and Ray Henderson's mechanics, did a won-derful job, as this compressed specification summary confirms:

> Reinforced body shell
> 15 in. wheels (13 in. were standard)
> Larger disc and drum brakes, calipers and drums
> 150 bhp engines with three dual-choke Weber carburettors
> Special wide-ratio gearboxes (to make o/d third and second more useful)
> Limited-slip differentials
>
> —and much extra detail.

Once again Fred Nicklin and Gordon Birtwistle re-developed the handling; considering their weight and bulk, the cars handled extremely well, and their traction (aided by the lim-ited-slip differentials) was excellent. They were much faster than standard, but much less economical, as a *Motor* road test later proved.

Four cars—AHP 424B, AHP 425B, AHP 426B and AHP 427B—were built up during the summer, and three of them were sent off to compete on the Spa-Sofia-Liege marathon. This was an incredible challenge for a new car. Even if we had been looking around, noth-ing tougher, dustier, rockier or more demanding could have been found. In those days it was even considered to be more difficult than the East African Safari.

Very few cars ever finished the 1960s-variety of marathons, and memory suggests that our ambitions were to get at least two cars back to Belgium, to beat the Rover 2000s, and to try to get at least one car into the top five or six places.

What turned out to be the last of the Yugoslavian-based marathons lived up to its reputa-tion, for only 21 of the 106 starters managed to finish. For Triumph, the good news was that all the Rovers and all the Fords retired   but the bad news was that all the Triumphs retired as well. No fewer than four of the well-developed Citroen DSs—with which the 2000 would have to compete in future events—made it back to Liege too.

All three cars reached Sofia, turned round after a ludicrously short one hour rest, and set off strongly for the Adriatic coast, and Italy. Half way through this run, close to Split, all three cars were well placed, but then suddenly broke down. All had the same problem—the rear suspension cross-beams pulled away from the floorpan of the body shells, and the cars could go no further.

The only consolation was that the failure was immediately diagnosed as a fatigue prob-lem. It was remarkable that all three cars had stopped within hours of each other—and the solution (to reinforce the body shells in that area) was immediately applied, with complete success, for future events. It was the first, and the last, time that a 2000 ever went out of an event with that problem.

Three Spitfire rally cars, ready to leave Coventry to take part in the Tour de France in 1964. This was the first time they had been fitted with GT6-style fastback tops. For this occasion only, the faired-headlamp bonnets had been borrowed from the Le Mans cars.

My lasting memory of that event (in the days when telecommunications in Yugoslavia were very sketchy) was of sitting in the sun at the Novi control, waiting . . . and waiting . . . and waiting. Alongside me were Ralph Nash of Rover, and Bill Barnett of Ford; we all spent a disappointing day in the sun, then eventually wondered how we could retrieve the cars which were clearly not coming.

Triumph needed a change of luck and, fortunately, the tide turned dramatically in September 1964. Not only did the Spitfires put up a remarkable show in the 10-day Tour de France, but they also astonished everyone with their performance in the Geneva rally.

The Tour de France had always been in our programme, and at class level it promised to be a real 'blood match' between the Spitfires, the Alpine-Renaults and the Bonnets. After the encouraging show at Le Mans I was confident that the Spitfires had the endurance, but would they have the pace?

Except that a lot of racing was involved, the Tour was almost completely different from Le Mans. For one thing, it made a complete 6,000 km circuit of France (and northern Italy—for a race was scheduled at the Monza circuit), with several one-hour long races at permanent circuits such as Rheims, Rouen, Le Mans, Cognac, Pau, Albi, Clermont Ferrand and Monza. There were also nights of high-speed road rallying in the Pyrenees and the French Alps.

Not only that, but the Tour de France was for homologated cars, which was both good news and bad news for Triumph. The good news was that Alpine-Renault would not be able to use the twin-cam engined space-frame 'specials' which had beaten the Spitfires at Le Mans—but neither would Triumph be able to use the lightweight Le Mans cars either.

For the Tour, therefore, three of the blue rally Spitfires were given slightly different versions of the GT6-style hardtops (with 'luggage lids' in the tail), we borrowed and repainted three of the faired-headlamp noses, we fitted the vast 18-gallon fuel tanks (the Alpine cars had used normal production-car tanks),—and we also fitted prototype all-synchromesh gearboxes, intended for use in the GT6 in due course.

The all-synchro gearbox, of course, was not homologated, but as we knew (and they knew that we knew) of major non-homologated items being used on cars used by our rally-

Jean-Jacques Thuner (standing inside the door of his Spitfire) paused for service on the Tour de France of 1964. Co-driver John Gretener (with moustache) is talking to Graham Robson, while Rob Slotemaker (far left) wonders what the problems are.

ing rivals, we decided to take the risk. In fact, Triumph was not 'found out' until mid-1965, as Gordon Birtwistle's narrative will make clear.

It took ten days for the Tour to go from Lille in the north-east of France to the finish in Nice, but after less than a day two of the Spitfires had already retired. The Bradley/Fidler car (ADU 6B) actually melted a piston on the warming-up lap of the first race at Rheims, while Valerie Pirie's machine threw a rod before the end of the hour. As *Autosport's* reporter commented: 'Roy Fidler was last seen heading towards the champagne tent with nine day's expenses.'

The other two crews then buckled down to their task, regularly beating the Renault-engined French cars in their class, and usually finishing well up on handicap. The handi-

Although the 'works' Spitfires were very hard-sprung, the rear wheels could begin to tuck in as the limit approached. This was Slotemaker being as brave as usual in the Tour de France.

Lining up for one of the hour-long races which were features of mid-1960s Tour de France events—here are two of the Spitfires. Car No. 131 was driven by Rob Slotemaker (his co-driver was Terry Hunter), while car No. 129 was driven by Jean-Jacques Thuner (co-driven by John Gretener).

cap? Too complex to explain, but all races were measured in distance covered and, depending on its engine size, a car had 'bonus' kilometres added to those achievements.

In a GT category which included shoals of Ferrari GTOs, Porsche 904s, Daytona Cobras and Alfa Romeos, the Spitfires (Slotemaker was usually slightly ahead of Thuner) continued to impress. As the Alpine-Renaults gradually fell by the wayside, Rob set third best on handicap in two hours at Le Mans, second in an hour at Cognac and was best of all at Albi and Clermont Ferrand.

Poor Thuner, who was always close behind, suffered from a blown engine in the French Alps, which left only Slotemaker, though he was well in the lead. Then it was heart-in-

Four highly-modified Triumph 2000s, all ready to leave Coventry to compete in the RAC Rally of 1964. All cars were white, with powder blue roof panels, and with anti-glare matt-black bonnets. The engines had three dual-choke Weber carburettors which developed 150 bhp, and special 15 in. road wheels were used. AHP 426B, driven by Roy Fidler and Don Grimshaw, finished sixth overall. (Gordon Birtwistle).

mouth time, for on the last day the Spitfire's fuel tank split a seam (fortunately near the top), which meant that on the last day there were fumes in the cockpit, the drivers both felt nauseous, and we wondered if the car would actually get to the finish.

In Nice, however, the Spitfire not only won its class, but finished fifth on handicap in the GT category and tenth overall in that category (behind four Ferraris, four Porsches and an Alfa Romeo). The blow to Alpine-Renault's prestige in France was catastrophic, and this was rubbed in a few weeks later when Thuner's car, having been re-engined, was entered in the Paris 1,000 km race at Montlhéery, where it won its class

Two of the rally Spitfires (ADU 6B and ADU 7B) were then re-prepared, and sent out to tackle the Geneva rally, which was Thuner's 'home' event, and very important to Triumph's marketing image in that country. because this was an event run on a handicap basis, it suited the fleet little Spitfires, which were already so quick that they sometimes set fastest GT category times on this event of tarmac stages and speed hill-climbs. After 2,000 km Terry Hunter not only finished second overall, but won the GT category, his class, and was one of the Triumphs which won the team prize. Thuner was fifth overall, and second in the class.

For the RAC rally, which ended the 1964 season, once again we relied on the Group 3 Triumph 2000s. This time all four of the cars were employed—the 'spare' being loaned to Peter Bolton, at Harry Webster's request—while the SMART Spitfire was once again prepared for Valerie Pirie to drive.

After the great successes of the Tour de France, the Geneva rally, and the Paris 1000 km race, 'works' morale was high, particularly as we thought that the 2000s were already a lot better than they had been in August. We were happy that the body shell breakages would never occur again—and we had also improved the traction considerably.

Dunlop had produced a new all-weather treaded radial ply tyre, the SP Weathermaster, which combined a knobbly tread with a high-performance carcase. because it sounded promising for off-road use, so in October we arranged to test at Bagshot (where Terry Hunter drove a 'hack' 2000, and I operated the stop watches). Back to back comparison of SP Weathermasters against ordinary (cross-ply) Weathermasters, and against smooth treaded SP3s, proved that the new tyres were several seconds quicker in a three minute loop. Instantly we tried to reserve as many as possible of the 15 in. variety, though BMC also reserved a large quantity of the same tyres to use on the Austin-Healey 3000s.

AHP 425B, driven on this occasion by Jean-Jacques Thuner and John Gretener in the 1964 RAC rally. The car looks immaculate (was this the very first stage of the event?), and is certainly riding high on its suspension.

In 1964 the RAC rally favoured brave Scandinavian drivers, and very fast cars (in that order), but we always thought the 2000s would be in the hunt. And so they were. Except that Jean-Jacques Thuner's car broke its rear axle gears, and Terry Hunter's car not only had a nibble at a tree, but also fell on to its side with not even a trace of body damage, the cars were extremely reliable.

Most of the magazine writers were thrilled by their character, and the noise they made. *Autocar's* reporter wrote:

> These cars. sounded superb, their easily distinguishable six-cylinder exhaust notes rising and falling as they echoed through the silent woods.

Well before half-distance the 'works' cars were in the top ten, and if Terry Hunter and his co-driver Patrick Lier had not overslept at the breakfast halt at Turnberry (and been penalized heavily for a late departure—even though they, and their car, were in *parc fermé* at the time—Terry would have taken fourth place. As it was, Roy Fidler took sixth place overall, the two modified saloons finishing immediately behind Timo Makinen's Austin-Healey, which won that class. In fact there was a real cliff-hanging finish in that class, for the Big Healey had split its gearbox casing in the Snetterton race circuit test near the finish, and only managed to limp back to London while laying a continuous trail of Castrol. If only . . .

That was the end of a busy, and extremely encouraging season—and with no change of rules, nor of rival cars, foreseen for 1965, Triumph looked ahead to the new year with confidence. All three sets of cars, suitably freshened up and developed, were to be retained.

Once again, driver changes were in the air. The existing team rally drivers were all retained for 1965—though to make sure that Terry Hunter didn't defect to other high-spending teams he was offered a contract which guaranteed him a fee for the whole of 1965.

Triumph's only big signing in the mid-1960s was when the young Simo Lampinen (who had just won the 1000 Lakes rally in a Saab 96), joined Triumph for 1965 and 1966. Here he signs his contract, supervised by Harry Webster.

There was also a new signing.

Simo Lampinen was a young Finn who had just won the Finnish 1000 Lakes rally in a Saab 96—beating Tom Trana's Volvo, Rauno Aaltonen's Saab and Timo Makinen's Mini-Cooper S, and I was delighted to persuade him to join the Standard-Triumph team for 1965. Part of the deal was that he would always be able to use left-hand-drive cars. This meant that two new rally cars—a 2000 (EHP 78C), and a Spitfire (AVC 654B)—were prepared during the winter, though as it transpired Simo did not actually use the 2000 until the end of the year

At first Harry Webster was reluctant to pay out about £2,000 a year (drivers were not as well-paid in those days) for an 'unknown' who limped badly because he had suffered from polio as a child—but a few demonstration laps at MIRA soon caused him to change his mind!

For 1965, in fact, we had too many drivers at our disposal—Lampinen, Hunter, Fidler, Slotemaker and Thuner—and I had to make it clear to them that not all of them would drive on all the events. Rob and Jean-Jacques, in fact, were quite happy to keep away from the rough-road events, though the other three were more ambitious!

1965

Ever since I had joined Standard-Triumph, I had spent most weekends co-driving in British rallies, usually in Ford cars. Then, for 1965, I was encouraged to be more loyal (that's a diplomatic way of saying that someone had noticed what was going on), so Roy Fidler and I schemed up a British programme around one of the Group 3 2000s.

This explains why Roy and I entered a car in the Welsh International rally in January, an event which we led, convincingly, until the engine began to misfire and progressively lost

For 1965 the rally Spitfires were fitted with this type of auxiliary headlamp installation—made fashionable by the arrival of single-filament quartz halogen bulbs which were used in the main headlamps. This was Valerie Pirie's car being prepared for the Monte Carlo rally of 1965.

Monte Carlo rally 1965—and this is not even on a special stage. Slotemaker and Taylor (in ADU 6B) struggling against the conditions in the French Alps. (Alan Taylor)

power. It took ages for our service crew to discover a distributor fault, by which time we were OTL. For Roy and I, this was the start of a frustrating season of British events, for we broke transmissions on several occasions, two of them while I was actually delivering the car to the start of the event.

For the Monte Carlo rally, we looked at the regulations, gambled with the weather, and guessed that nimble Spitfires would be more competitive than big 2000s. Three team cars were entered—including the new left-hand-drive car for Simo Lampinen—while once again a 2000 was loaned to Peter Bolton; the 2000, in fact, disappeared at an early stage with a blown engine. The Spitfires received their definite 'face-lift', with the addition of inboard auxiliary headlamps (and the use of single-filament quartz halogen bulbs in the normal headlamps).

Rob Slotemaker (in jumper) and Alan Taylor (in anorak), shared the Spitfire ADU 6B in the Monte Carlo rally of 1965, finishing second in class in a very wintry event. This is Monte Carlo, where it never snows! (Alan Taylor)

ADU 6B, driven by Rob Slotemaker and Alan Taylor, trickling through a French town on the long run down to Monte Carlo in the 1965 event.

Terry Hunter celebrated his first fee-paying drive by over-indulging on the night before the start in London—but worse was to come. As in 1963, the weather was appalling, and although Lampinen and Slotemaker reached Monaco unscathed, Hunter crashed his car and was forced out of time.

At the start of the final Mountain Circuit (which included six long special stages) Lampinen was in tenth place, with Slotemaker not far behind (there was a handicap to confuse the casual observers). During the night Lampinen's car blew its engine, and Rob Slotemaker eventually finished second in his class—the winner of that class being Timo Makinen (Mini-Cooper S) who also won the rally outright. The gamble with the weather had failed.

Terry Hunter, having reached Monaco in his rally car, then had another celebratory evening, managed to get on to some of the special stages before the rally arrived, was

Slotemaker's Spitfire nearing the end of the Monte Carlo rally of 1965—at the control at Pont Charles Albert, north of Nice.

Bob Tullius driving a Spitfire race car at Sebring in March 1965. Note the position of the brackets for quick-lift jacks, and the massive filler cap for the 18 gallon fuel tank.

promptly arrested by the French police, and spent the rest of the night in jail. The result was that he was immediately banished from the team, and was never employed by another organization. This was a great personal tragedy for him, as when he was in the appropriate mood he was a very brave driver indeed.

This was the moment when I left the Standard-Triumph team to pursue a career in motoring journalism, though I was very happy indeed to hand over to the joint management team of Ray Henderson and Gordon Birtwistle, both of whom had a lot of experience. In the next year or so, Ray looked after policy and preparation, while Gordon looked after office work, and carried out all the dynamic testing. Any doubts that I had for deciding to leave Standard-Triumph were banished by the events surrounding the Monte Carlo rally.

Even before I moved out, however, I had been involved in the plan to send a team of further-developed race cars to compete in the Sebring 12-Hour race. In March, when the event took place, Ray Henderson and 'Kas' Kastner ran the operation between them.

For Sebring, the aluminium-GRP bodied Spitfire race cars were equipped with GT6-style gearboxes, and entered as homologated GT cars, which was cheeky as in theory the current Appendix J regulations only allowed body *skin* panel materials to be changed. However, since the 'works' MG Midget also sported a very non-standard looking shell, no-one complained. As Gordon Birtwistle quipped: 'Maybe the scrutineers didn't have any magnets at Sebring!'

This particular Sebring race was 'the year of the cloudburst' yet only one of the cars ended up off the road, the others finishing closely behind the Midget in its class. That was the one and only time that a Spridget ever beat a Spitfire in a straight fight.

Since Peter Bolton rolled ADU 1B—it had only just been re-created after the Le Mans crash—there was a lot of work to be done before Le Mans in June, especially as Harry Webster's engineers had plans to take weight out of the cars in that time.

After the blizzard conditions of the Monte Carlo rally, the rally team was looking forward to balmy spring-like weather on the Tulip rally, especially as its route went as far south as Switzerland in May; once again they were thwarted, for it snowed heavily in the French mountains surrounding Geneva.

As before there were handicaps to be considered, which explains why two Group 3 2000s, and a single Spitfire were entered. Early in the event Lampinen's Spitfire devoured its clutch (a rare failing on this little car—though the high first gear, and repeated hill starts on the speed hill-climbs, can't have helped) but the 2000s boomed along, their traction

Sebring 1965, where three Spitfire race cars started, and two finished strongly, second and third in their class. ADU 3B made the journey, but was not used on this occasion. (Mike Cook)

Triumph and MG competed against each other many times, with honours broadly even over the years. This is Sebring, March 1965, where the Bolton/Rothschild Spitfire is battling with one of the special-bodied MG Midgets. (Mike Cook)

Jean-Jacques Thuner (right) and John Gretener (left) drove this left-hand-drive Triumph 2000 to a class win in the Tulip rally of 1965. Gordon Birtwistle (centre) looks pleased with the result. (Gordon Birtwistle)

being a real advantage in the slush and snow; even so, on the Col de la Faucille, Slotemaker needed to use car floor mats to give his rear wheels enough traction at one point; eventually the engine failed.

Jean-Jacques Thuner (in the new left-hand-drive car) drove extremely well to finish third in the GT category—only Rosemary Smith and 'Tiny' Lewis (Hillman Rallye Imps) finished ahead of him, thanks to the handicap—and won his class. *Motor* borrowed the car for a performance test immediately afterwards, when it had been fitted with a milder camshaft profile, Roger Bell noting 0-100 mph in 33.9 sec., but overall fuel consumption of an appalling 12.7 mpg.

Jean-Jacques Thuner and John Gretener about to start their 'home' event, the Geneva rally, in June 1965. Driving ADU 5B as usual, they finished fifth overall and won their class. (Gordon Birtwistle)

Roger also described the habitual use of 6,500 rpm in the gears, an ability to pass Ferrari drivers on the M1, mentioned that the car was : 'capable of leaving long black lines on the road as it gathers speed from standstill, and which can rend the air with a noise like hoarse thunder . . . [with] negligible body roll and tremendous adhesion inspired hard cornering.'

Next month was Geneva rally time, an event in which the Spitfires had been so success-ful in 1964. Since it was Thuner's 'home' event, the factory once again entered two cars, and looked forward to another good result.

Until, that is, the cars were presented at scrutineering. Gordon Birtwistle now takes up the story:

> Someone had talked. The scrutineer wanted to know how many gears were synchronized, and I told him the vehicle was in accordance with homologation papers. That was a porky-pie . . . .
>
> He insisted on driving the car. I told him that no-one but my drivers and myself would drive the cars—but we were in a tricky situation, especially as the Le Mans cars had the same gearbox fitted. I went off to have a word with Jean-Jacques then, suddenly, I heard the Spitfire started up in the hall, and the scrutineer reversed at high speed, then rammed it into first gear and acceler-ated forward. *That* change was as smooth as silk.
>
> The scrutineer then gave us an ultimatum. We could only start the event if we rebuilt the gearboxes and, if we won anything, we had to have them stripped afterwards.
>
> There was no choice. We took the cars to Blanc & Paiche, the Triumph importers, borrowed two cars from the showroom, worked through the night to install non-synchromesh first gears, and ran them like that.

There was a happy end to the story, for the two rebuilt cars went out on the event and took first and second places in their class, with Thuner also finishing fifth overall. The miracle is that there were no long-term effects of this little farrago, and word never got as far as Le Mans.

All four of the original 'ADU' Spitfire race cars were prepared for Le Mans in 1965, where they were entered as homologated GT cars, though the team only expected to see three of them starting the event. On this occasion they were about 110 lb. lighter than in

This informal team picture was taken at Le Mans in June 1965. Left to right: Claude Dubois, Peter Bolton, Roy Fidler, Rob Slotemaker, Jean-Jacques Thuner, Simo Lampinen, Bill Bradley, Sir Donald Stokes, (unknown) and George Turnbull. (Maxwell Boyd)

1964. One reason for the saving was the use of the smaller gearbox, and smaller rear brakes, but there was also a thinner-gauge chassis frame (which saved 31 lb.), and for the first time the race cars used the aluminium cylinder heads which had previously been exclusive to the rally cars.

For the fourth (reserve) car, John Lloyd and Ray Henderson called up Jean-Jacques Thuner and Simo Lampinen from the rally team, while Roy Fidler was also contacted:

> Ray Henderson asked if I fancied myself as a racing driver. I didn't, but they nominated me as reserve Pilote anyway. I had to qualify, of course. Going down the Mulsanne straight, the Fords and Ferraris passed me as if I was standing still, and the whole car was blown sideways. Driving one of the slowest cars in the race—well, it was bloody dangerous; you had to be looking behind you all the time.
>
> Simo, Jean-Jacques and I, we thought it was just possible to take White House corner flat out, given a clear road. That's all very well, except that the second you were committed to a line you suddenly got a Ferrari up your backside.

As it happened, all four cars started the event, and on this occasion two of them completed the 24 hours. In every way this was a better performance than in 1964, for the cars finished 13th and 14th and were first and second in their class, which gave the factory *and* the French importers something to shout about. Once again, the opposition from Alpine-Renault was obliterated, for not a single car finished.

The cars were slightly more powerful (109 bhp instead of 98 bhp) and faster than in 1964—about 200 rpm (or perhaps 4 mph) in a straight line—but this was enough to cause vibrations in the oil coolers. The Bradley/Bolton car retired after all the oil had leaked away and the engine blew, but Slotemaker's demise was more personal—he lined up for White House, pulled over to let a faster car through, dipped his lights, and ran out of road.

This, in fact, was the last 'works' appearance by the Spitfire race cars, though some of them were shortly lent to a private team to compete in the Six Hour Relay race at Silverstone, and Bill Bradley continued to race a car in British and European sports car events until the end of the season. The rest of the year, therefore, was the province of the rally team.

This 1965 Le Mans pitstop study shows how the Spitfire race cars had changed in detail since their first appearance, for there was now a shield through which the carburettor intakes protrude (to keep intake air cool), and there was a special cool air duct which kept the interior of the car cool.

Look carefully at the registration number of this car, and wonder whether this really was ADU 6B, or another of the 'works' rally cars. In any case, this is Roy Fidler driving in the Six Hour Relay race, Silverstone, in the summer of 1965. (Roy Fidler)

Roy Fidler's frustrating 1965 season continued in June, in the Gulf London rally. Once again I was his co-driver in a high-speed 'British forestry marathon' (the Gulf London organizers made sure that road sections were equally as demanding as those of the stages), once again he led the event outright, and once again the car let him down. It was a familiar story—the 150 bhp had proved too much for the transmission, which broke a half-shaft in the 30-mile Dovey stage. As this event currently carried Britain's top rallying prize (£1,000) both Roy and I were devastated.

Then came the French Alpine rally, and what was really the Spitfire's finest hour. On this occasion the organizers, in their wisdom, arranged a category for 'Sports and Prototype cars', which gave every red-blooded manufacturer the chance to build its best-possible rally car without homologation rules getting in the way. The organizers also set the most difficult time schedules yet seen on this high summer/mountain race.

All four of the fast-back rally cars were prepared (ADU 8B, the bubble-top spare, was

Gordon Birtwistle (with homologation papers) discussing a tricky point with Simo Lampinen, at scrutineering for the Geneva rally of 1965. At this moment there was a big problem—the gearbox had just been declared illegal! (Gordon Birtwistle)

'Kas' Kastner (right) inspecting one of the 1.3-litre Spitfire engines which would be used in the Spitfire rally cars in the 1965 Alpine rally. Engine workshop foreman Ted Silver is in the middle, and technician Dennis Barbet is with them. ('Kas' Kastner)

never actually used on an event, but kept for practice only), and were aimed at the proto-type category. To take advantage of this, Harry Webster's engine development technicians produced four unique big-bore 1.3-litre versions of the highly-tuned engine (1,296 cc instead of 1,147 cc) this being an engine size which was about to be launched in the front-wheel drive 1300 saloon. With very little drama, the 70X units were turned into 79X units—the number being the displacement of the engines in cubic inches.

Rob Slotemaker and Alan Taylor on the Alpine rally of 1965, where the Spitfires used prototype 1.3-litre engines.

Rob Slotemaker's 1.3-litre engined 'prototype' Spitfire on the Alpine rally of 1965. The Spitfires dominated this category, finishing first (Lampinen) and second (Thuner) overall, their best-ever result.

These engines produced no less than 117 bhp at 7,000 rpm, and 97 lb.ft. of torque at 5,500 rpm, which made them a lot more flexible, with more mid-range punch. To match this the cars were also fitted with the 1964 Le Mans-type of braking, which included 9 in. rear brake drums.

Ray Henderson was kind enough to ask me along to co-drive for Roy Fidler, on what was to be my first, and only, 'works' ride in the car which I had helped to create. The event

Simo Lampinen, in his left-hand-drive Spitfire, climbing a bleak Alpine pass in the 1965 event, where his car won the entire prototype category. (John Davenport)

There is a big drop over the edge of that road in the French mountains. That is John Sprinzel and Willy Cave competing in the Alpine rally of 1965. (John Sprinzel)

started well for us, and we were settling down well, and looking for a clean run on the road when, as Roy Fidler recalls:

> After about 12 hours we were hurtling along. I think we'd just had the wheels changed, and there was a drop on the left side and a big mountain on the right. All of a sudden I saw a rear wheel passing us, it bounced once, then disappeared, never to be seen again by a human being. After that I couldn't steer the car, it slithered along, and went straight across a slight-right-slight-left before I managed to stop it.

We were very lucky not to suffer any damage, for the rear wheel studs had broken, and that was the end of our rally. It was the only time that I ever had to abandon an event and fly home still wearing my racing overalls!

At the start, the Prototype category included six-cylinder engined Porsche 904s and Matra-Bonnets, but after three long sectors, and a climb of seemingly every high pass in France, they had all fallen by the wayside. Even for the Spitfires the schedules were impossibly tight, but at the end of the day Simo Lampinen and Jean-Jacques Thuner won the category outright. Apart from our broken-wheel-stud story, Slotemaker's car blew its engine, while the 2000 which was loaned to John Sprinzel broke its gearbox (but that was hardly unexpected).

In the next few weeks the race car mechanics then completed a unique Spitfire racing car, which was built and exported to the order of Walter Sulke, who ran ZF Garages, the Standard-Triumph distributorship in Hong Kong. This car, known then (and later) as the 'Macau car', was based on the layout of the race cars, but was different in many ways.

This machine was not a conversion of one of the Le Mans cars, but was built up from the stock of parts and panels. Visually there was a touch of Jaguar D-Type about it, for it was an open car with a moulded headrest behind the driver's seat, with a metal spine between the seats, and with a rigid cover over the space for the passenger's seat. It was on ultra-wide (5.5 in. rim) alloy wheels, had an extra-large petrol tank of 22.5 gallons, and the latest 109 bhp 1.1-litre engine. The top speed was over 130 mph, even with a 4.1:1 axle ratio, and it lapped the MIRA banked circuit at 122 mph.

After a short and varied career in Hong Kong, this car was shipped over to 'Kas' Kastner in Los Angeles. Kas told me that:

> I immediately put a 2-litre 6-cylinder GT6 engine into it, with three Weber carburettors. We had about 220 bhp and the car ran in SCCA D-Modified events. It was reasonably competitive—in this form it had tremendous straight line speed, but it didn't handle well, because we couldn't get big enough wheels and tyres inside the body; but we had a lot of fun with it.

This page In the summer of 1965 Triumph built up a very special Spitfire race car, to be sent to race at Macau (in the Far East), and other events in the same area.

By 1968 the Spitfire 'Macau car' had found its way to California, where 'Kas' Kastner's team installed a 2-litre six-cylinder GT6-style engine, a TR6 gearbox, and raced it in USA SCCA events. ('Kas' Kastner)

The RAC rally was the last big event in the 'works' team's programme for 1965, with no fewer than four of the 'works' 2000s starting the event—one of them being loaned to John Sprinzel, whose co-driver on this occasion was David Benson, the motoring correspondent of the *Daily Express*. On this occasion Roy Fidler's car ran in slower, Group 2, guise; the others being familiar, in Group 3. Ray Henderson was planning ahead, and knew that his Weber-carb'd cars would be outlawed by new rules in 1966.

   This was the rally which nearly ground to a halt in Yorkshire, in Scotland, and again in Wales, where heavy snow, then a thaw, followed by a severe frost made stage driving conditions almost impossible. It was a dramatic end to this phase of Appendix J rallying, for there were 162 starters ready to tackle 57 special stages.

Roy Fidler and Alan Taylor at the start of the 1964 RAC rally. They went on to finish sixth overall, and second in their class. (Gordon Birtwistle)

This page On the RAC rally of 1965, where it was sometimes icy and always slippery, Simo Lampinen seemed to have his Triumph 2000 sideways for most of the time. Simo led the rally at one point, until his engine blew.

Simo Lampinen's left-hand-drive car started from No. 6, the others being well spaced at Nos. 23, 51 and 54—which made life difficult for Ray Henderson's service crews. Half way through the event, by which time 76 cars had already retired, Lampinen was up to fourth place overall, with Roy Fidler close behind, and leading his class.

By the time the cars struggled out of the Yorkshire blizzard, Simo's engine was failing, and shortly expired (a blown gasket was diagnosed), but Fidler's near-standard 2000 was healthy, was much faster than the Rover 2000s in its class, and was seventh overall. Two days later, at the end of the event in London, the 2000 had improved to fifth place, and Roy took the 'Best British driver' award. Jean-Jacques Thuner finished second in his class—as usual to an Austin-Healey 3000—and Triumph were highly delighted, particularly as they had once again humiliated their bitter rivals—the Rover 2000s.

Once again, I was closely involved, in the last event of the season, where Simo Lampinen and Roy Fidler both used their ex-RAC rally cars (which had been speedily refurbished

Roy Fidler and Alan Taylor won their class, finished fifth overall, and took the 'Best British' award in the RAC rally of 1965, using one of the well-known quartet of 'works' 2000s. (Alan Taylor)

Gordon Birtwistle (in 'natty 'at') and Roy Fidler on the RAC rally of 1965. (Alan Taylor)

back in Coventry) to compete in the Welsh International rally. My connection was that Ford had invited me to co-drive for Roger Clark in his first outing in a 'works' Lotus-Cortina.

This, in fact, was the second Welsh rally of 1965 (the first having been held in January), and was a two-day blast around the Principality, using many of the same stages as the recent RAC rally. Throughout the event it was a straight fight between one Ford—that of Roger Clark—and the two Triumphs, driven by Lampinen and Fidler.

After 12 hours Clark's Lotus-Cortina was only nine seconds ahead of Lampinen. Lampinen was in the lead after the cars had raced through two Dovey stages, but both cars took a pounding and needed attention to their suspension and steering gear during the day. Llanbed forest sorted out a result, when Lampinen went off in thick fog for ten minutes, the result being that the 2000s had to settle for second and third places, overall. Even so, this was their best-ever performance at international level so far; if only they had had a future in Group 3 form.

Well before the RAC rally had been run, however, Ray Henderson and Gordon Birtwistle realized that their Spitfires and 2000s were both coming to the end of their careers, well before they had reached their peak. Major changes to Appendix J were brewing, and would come into effect from 1 January 1966.

For Triumph the effects would be devastating. On the one hand it would no longer be possible to run much-modified touring cars in Group 3 (which killed off the Weber-carburated, 15-inch wheel 2000s at a stroke). Faced with competition from Ford's Lotus-Cortina and BMC's Mini-Cooper S, it looked as if the 2000 would no longer be competitive.

On the other hand the use of alternative cylinder heads, different body styles and lightweight body skin materials for Group 3 cars was also to be outlawed, which immediately meant that the Spitfire would also become uncompetitive, not only for rallying, but for the Le Mans 24 Hour race; this was a real body blow.

Look at what might have been possible, if there had been no big changes, and if Harry Webster could have gained permission to forge ahead? The Spitfires could have been homologated with 1.3-litre engines, and engine reliability would certainly have improved in 1966, while the GT6 version, complete with a race-tuned engine would have followed for 1967. The 2000s could have adopted the fuel-injection which was already being developed for the TR5 sports car, and there was a lot of weight that could have been taken out, by the use of aluminium panels.

Triumph, incidentally, was not the only marque to suffer. BMC lost the performance edge of its Austin-Healey 3000s, while Porsche and Alfa Romeo both suffered by losing special-bodied cars. But it wasn't all bad news—those teams which benefited were BMC (with the Mini-Cooper S), Ford (with the Lotus-Cortina), Porsche with the 911, and Alfa Romeo with the specially-homologated Giulia GTA.

For its 1966 programme, therefore, Standard-Triumph was faced with big decisions. Should it develop a homologation special? Should it make the best of existing cars? Or should it withdraw from the sport which had served it so well in recent years?

Chapter 6

Late 1960s sabbatical

1966-1969

Even before Ray Henderson could settle his programme for 1966, the Monte Carlo rally organizers threw a spanner into the works. As I wrote (for Peter Garnier) in *Autocar* in August 1965: 'In order to encourage private entrants, the handicap system would favour the Group 1 cars. . . .' The portents were ominous.

Then, in October, I was able to confirm that for 1966: 'The complicated factor of comparison used in past years has now been dropped. All Group 1 cars will be scratch, and those in Groups 2 and 3 will have a coefficient of 1.18 applied to their scores. . . .' Later we noted that for a Group 2 or Group 3 car to match the penalty of a Group 1 car which took 10 minutes on a special stage, it would have to beat 8 min. 29 sec., which was clearly impossible.

It was the first time the Monte Carlo rally planners had leaned so heavily towards the use of 'standard' cars on its rally, and this had almost certainly happened because of pressure from French manufacturers. In the days when a Monte could be won by accurate time-keeping, a Citroen or a Renault could win, but from 1961, with special stages to sort out a result, and with ultra-fast 'homologation specials' taking the awards, the balance had shifted.

Was it an attempt (like 1961) to give victory back to the French? We were about to find out.

Ray Henderson and Gordon Birtwistle had great difficulty persuading Sir Donald Stokes to back an entry in this event, for it was clear, right from the start, that Group 1 (i.e. standard-specification) Triumph 2000s would not be competitive. All that could be hoped for was an honourable finish and maybe, just maybe, the chance of taking a team prize.

1966

As Gordon Birtwistle recalls:

> Sir Donald Stokes was only interested in competing with a winning package. But with twin Stromberg carburettors the 2000 wasn't competitive. . . .

So no-one had any illusions, but somehow Harry Webster preserved a budget for 1966—there were two interesting race car programmes in prospect—and approved a Monte Carlo rally entry of three cars. 'Comps' actually prepared a quartet of four new cars, all of them in pure white, with overdrive transmission. One of those cars—FHP 991C—was only used in practice.

Fleet Street journalists Tommy Wisdom and Courtenay Edwards, along with Jeff Uren, also started the event in a maroon coloured 2000, which was registered FHP 989C, though this was not an official 'works' car. Instead it was one of the Press fleet, which was privately prepared in the service department, and was not even provided with service support from the 'works' team.

Even at the start of the 1966 Monte Carlo rally, Tommy Wisdom already has his monocle in place. Although this Triumph 2000 carries a 'works' (Coventry) registration number, it was actually prepared by the service department, not Ray Henderson's competitions shop. (Gordon Birtwistle)

You wouldn't think so, but this is a scene on a winter rally—actually the Monte Carlo of 1966, where Roy Fidler and Alan Taylor are tackling one of the road sections (no crash helmets!) in their Group 1 Triumph 2000. (Roy Fidler)

Preparation details of the Group 1 Triumph 2000s of 1966 included only two extra Lucas fog lamps (regulations!), Dunlop Weathermaster tyres, and stick-on number plates (which the British police rather frowned upon! This shot was taken on arrival in Monte Carlo. (Roy Fidler)

Compared with the well-honed 1965 'Group 3' cars, the 1966 Monte cars were heavy, less powerful, and slow. They had to run with bumpers, they had to use ordinary 13 in. wheels, and they produced very little more than the catalogued 90 bhp. One useful 'tweak', which was naturally never publicized, was that overdrive was made to work on second, as well as third and top gears. The drivers remember this as an absolute blessing, as this filled in a sizeable 'hole' between second and third gears, and allowed the car to keep flowing along at peak torque the whole time.

As it happened, the event was dominated by a disqualification scandal involving the BMC Mini-Cooper S cars, and the 'works' Lotus-Cortinas. 'On the road' the Minis won easily—taking the first three places—but the organizers could not believe that such tiny cars could possibly have set the fastest times on the stages. Ruthless scrutineering—of *every* part—eventually inspired the officials to disqualify these cars on the grounds that their modified headlamp arrangements were 'illegal'. Need I remind anyone that as a result of this charade it *was* a

Roy Fidler and Alan Taylor in action, on a special stage of the 'Group 1' Monte Carlo rally of 1966. (Alan Taylor)

In December 1965 Triumph's GT6R Le Mans race car project had got this far. Basically it was intended to use the chassis frame and body shell of a 1965 Le Mans Spitfire, but a highly-tuned 6-cylinder 2.0-litre engine. Although fuel-injected installations were already being developed for this engine, the only GT6R car to be partly built had three dual-choke Weber carburettors of the same type as those used on the 1964-1965 Triumph 2000 rally cars.

French car—a Citroen in the same 2.5-litre class as the Triumphs—which 'won' the event?

None of this affected the performance of the 2000s, which were running with their normal twin headlamps, and just two extra Lucas driving lamps. They were, however, outclassed by the 'works' Citroens and 'works' Lancia Flavias, which took the top four places in the rally. All had front-wheel-drive.

Of the three 'works' 2000s, only one—that of Roy Fidler and Alan Taylor—finished the event, taking a creditable 14th place overall. Incidentally, it was also fourth overall in the unofficial 'front engine/rear drive category'—a great showing on this wintry event where to win, a car really needed to have its driven wheels and its engine at the same end of the car!

Both the other 'works' cars broke their transmissions—Lampinen's destroying its gearbox, and Thuner's car its back axle—while the ebullient Fidler found time to have a faulty dynamo changed, along with a battery which had gone flat, and slid off the road at one point and damaged the rear corner of his car.

The sleazy outcome of the 1966 Monte Carlo rally caused several manufacturers to look hard at their involvement in motorsport. Was it worth continuing? That rally, and others, seemed to be on the wane. *Autocar's* leading article of 28 January 1966 summed up thus:

> Virtually established and maintained by the British, this fine event has just begun to wane in popularity; the uproar over the results this year could mark the beginning of the end.

Elsewhere in the report, Peter Garnier commented that:

> As someone said: 'That's the last nail in the coffin of an already dying event.' One hoped, in the circumstances, that perhaps it might be.

Almost immediately after the shambolic Monte Carlo, Rover announced that it was abandoning its motorsport programme, while Triumph discreetly started to run down, but not close, its own operations. In future there would be no 'works' rally team operating out of Coventry, though help would be given wherever possible. In more ways than one this was a shock to the drivers, as there had been no warning. If the department was to be closed, they reasoned, why had a set of new cars been built up for the Monte Carlo rally?

Nevertheless, Simo Lampinen and Jean-Jacques Thuner were both released from their agreements, while Roy Fidler arranged to buy his ex-Monte car when it had been repaired, so that he could contest the RAC British rally championship for the rest of the season. For this programme, which included an earlier type car as well, he got a lot of material help from the factory, though this was often of an 'unofficial' nature.

Even so, there was still a lot of activity at Fletch. North. Roy Fidler's prize for winning his class in the 1965 RAC Rally was a free entry in the Swedish. So he took that up, using his 1965 car (ADU 426B) in less-modified (Group 2) form, with 13 in. wheels and less power. However, in what was a bitterly cold event, run in a lot of snow, he spent much of the time teetering on the edge of an accident, and finally had a good one which marooned his car until he ran out of time. His ex-teammate, Simo Lampinen, finished second in a front-wheel-drive Saab.

GT6R—the 1966 Le Mans project

By this time an intriguing race car project—the combination of Le Mans Spitfire and prototype GT6 engineering—had also been abandoned. The first car, which had never been completed, was broken up, and no trace remains, but it is worth detailing its history.

After Le Mans in 1965 it was clear that the Spitfire could go no faster unless it had more power. In any case the pending Appendix J rule changes meant that it would also be a heavier car in 1966, and that it would not be able to use the special fastback body style, or the 'E-Type' recessed-headlamp bonnet. Nor was there any chance of the 1.3-litre engined car being homologated until 1967 (when the Spitfire Mk 3 would finally go on sale).

The publicity to be gained from a sturdy performance at Le Mans was considerable, however. So in 1965, with completion in time to compete at Le Mans 1966 in mind, Harry Webster asked Ray Henderson to start building a new prototype race car—which, for obvious internal reasons, was dubbed GT6R.

Work on the GT6R 6-cylinder Le Mans race car project only progressed this far by the end of 1965. From February 1966, when Triumph's motorsport activity was rapidly wound down, the project was cancelled. One particular detail, not seen on the Le Mans Spitfires, was the pedal box/balance bar type of split hydraulic braking system.

Gordon Birtwistle confirmed that this car was never completed, but that the object was to run a 2-litre car in the prototype category at Le Mans, where it would have to compete against the Porsches. The basis of the design was the Le Mans Spitfire, complete with aluminium panels, the faired-headlamp nose and the fastback coupé roof, but the chassis design was new; this was the first competition Spitfire-based car to use a Lotus Elan-type/MacPherson strut type of independent rear suspension.

This suspension was mounted on to an extra pressed and fabricated cradle which was welded to an existing chassis frame. There was no question of the top of the strut being fixed to the body shell, which was certainly not strong enough to withstand the loads.

Power was by courtesy of a highly-tuned 2-litre six-cylinder engine of Triumph 2000 type, though it had the new type of cylinder head casting which would eventually feature in the Triumph TR5 and later GT6s. There were three Weber carburettors and peak power was expected to be 175 bhp. Lucas fuel injection, already tried on a Le Mans Spitfire, was proposed when it had become reliable, but was never actually fitted to this car.

Backing this engine was a TR4/2000 type of gearbox and overdrive, and a TR4A/2000 type of chassis-mounted final drive, which proves that good old-fashioned 'parts bin' engineering was alive and well at Fletch. North. The first time this power train had appeared in a competition car was in 1963, in the Spa-Sofia-Liege 'special' driven by Vic Elford, while the 1964 Le Mans cars had also used a TR4 type of gearbox, but no overdrive!

The 2-litre car would also have had larger brakes, wheels and tyres, and it was hoped that it would have a top speed of between 150-160 mph. Photographs printed in this book show that the car reached the lash-up/rolling-chassis stage by the beginning of 1966, but shortly after this the project fizzled out, and it was eventually broken up.

Why? There were several reasons. One was that the engine was by no means powerful enough or reliable enough for what Harry Webster had in mind—Bill Bradley's saloon car race programme proved that—and another was that this meant that the car's top speed at Le Mans would put it a long way behind the Porsches. In 1965 the fastest Porsches had exceed 160 mph, and would certainly be faster still in 1966.

The clincher, I am sure, was that Sir Donald Stokes was gradually, but inexorably, turning away from a motorsport programme, and vetoed the programme before too much time and money had been spent. All was not lost, however, for the rear suspension would be valuable in Bill Bradley's racing Spitfire programme, and *any* experience with the new six-cylinder engine was thought to be worthwhile.

Rallying in 1966

Once the decision to run down the rally team had been taken, the cars were quickly dispersed, and only two cars were retained. Both were ex-Monte Carlo '66 cars. One of them (FHP 993C) was sold to Roy Fidler—and another was 'de-registered' and earmarked for a race car programme in the British Saloon Car Championship.

Before Roy took over his 1966 car it was fitted with many items previously used on the earlier 'black bonnet' Group 3 2000s, most particularly the triple Weber engine, and the better gear ratios—but not the special 15 in. wheels. Roy remembers that the handling was not nearly as good on 13 in. wheels as it had been on 15 in. wheels. Maybe that was because there was not such a good choice of tyres in the smaller size but:

> The handling was certainly very different on the 1966 car. The early Group 3 cars handled very well—they were great big tanks but they handled well. Perhaps it was because the later car was a lot heavier. Certainly it was OK in the dry, but as soon as it rained well, on tarmac it was just as if I had great cobs of butter instead of tyres on the corners!

On the all-tarmac Circuit of Ireland Roy took fourth place—behind Tony Fall's winning Mini, Brian Melia's Lotus-Cortina and Adrian Boyd's Cortina GT—but on the Gulf London he retired his car when the rear suspension failed, after which the engine blew a

Roy Fidler and Alan Taylor in the ex-works Triumph 2000 which brought them so much success in 1966. This was the car Roy used to win the RAC British Rally Championship in that year. (Alan Taylor)

piston. He had been in third place overall at the time.

Then came the RAC International rally where the Fidler/Taylor crew, backed by a privately-entered car for John Sprinzel/David Benson, campaigned the big 2000s. Both cars suffered transmission problems early on—Fidler's car suffering from rear axle breakdown (nothing new in that!), and Sprinzel having gear selection troubles—Sprinzel eventually retiring with a blown gearbox and Fidler with a blown engine.

Fortunately for Roy, who had had a difficult year, it all came right in the end. Earlier in the season he had already notched up second place in the Express & Star national event, and an outright win in the Bournemouth. Then, in December, immediately after the end of the RAC rally, he took second place overall in the Welsh International, to become the RAC British rally champion driver for the year.

In those days there was no prize money, not even a trophy from the RAC, for winning their Championship, but the Standard-Triumph factory was delighted to make up for that. Roy Fidler tells me that Harry Webster asked him to visit the factory and that:

> As I walked in, he had organized a Press conference, where he presented me with a silver cigarette box, which had engraved on it:

> 'Presented to Roy Fidler by the directors of Standard-Triumph International Ltd in recognition of his brilliant achievement in winning the 1966 RAC British Rally Championship in a Triumph 2000.'

> — by the way, there was a cheque for £100 in it!

Racing in 1966

Although the scheme to take GT6Rs to Le Mans came to nothing, Ray Henderson's mechanics found plenty to keep them busy on the race tracks.

Once again Bill Bradley put together a Spitfire racing programme, whereby his car was prepared—and, as often as not, repaired—at the factory, but he actually paid for all the running expenses.

Triumph's 1966 2000 race car programme, with Bill Bradley driving this fuel-injected Triumph 2000, was not a success. The car was an ex-Monte Carlo rally 1966 machine. The author (right, in suit) is awaiting his turn to drive the car in a test session at Brands Hatch. (Gordon Birtwistle)

For 1966 this race car, which carried the number ADU 2B, was given the strut rear suspension which had originally been intended for the GT6R project and, as Bradley himself once told me in a letter:

> During the latter part of 1965 a lot of testing was done, and during the winter of 1965-6 Standard-Triumph agreed to run a special Spitfire for the 1966 season. A lengthy list of modifications was agreed to by Harry Webster. This resulted in easily the fastest and best handling Triumph Spitfire that was ever produced.

In fact some of that track testing had been carried out by Mike Costin of Cosworth, though Cosworth Engineering was never involved in work on the engine.

This car was unbeatable in its class at first, but at the Nürburgring it collided with the car which was leading the race, when leading both 'works' Austin-Healey Sprites by several minutes, and was effectively destroyed. The replacement car, somehow, was never quite as effective.

Nevertheless, with the aid of the strut-suspended chassis, Bill Bradley won 14 classes out of 18 race starts, and the only time he had to retire was as a result of the Nürburgring crash. At the end of the season this car held class lap records at Goodwood, Brands Hatch, Crystal Palace, Snetterton, Oulton Park and Mallory Park.

At the end of 1966 the last evidence of Spitfires in racing and rallying had been removed from the department. As Ray Henderson once told me, some of the cars were gradually 'used up', not only by Bill Bradley's programme, but by the work which went into the GT6R project. The surviving cars were all sold off, and as recently as the late 1980s one was recovered from oblivion in the south of France!

The other ambitious race programme in 1966 was to develop a Triumph 2000 for Bill Bradley to drive in the British Saloon Car Championship, where a new set of regulations allowed what were called Group 5 cars to compete. Effectively these were Group 2 cars with a mass of extra tune-up equipment added, for engine induction, brakes and transmission modifications were all allowed, as were extra-wide wheels. As with the old rallying rules, Standard-Triumph could see that if the engine regulations were virtually thrown away, then the 2000 could be much more competitive with the Lotus-Cortinas and—as it transpired—

the lightweight 1964-style Monte Carlo rally Ford Falcon Sprints.

Bill Bradley put in some of his finance, but the development of the car was all done at Fletch. North. One of the ex-Monte Carlo 'Group 1' cars was stripped out and re-prepared. The 2-litre engine was given a prototype Lucas fuel injection installation (of the type which was being developed for the forthcoming TR5 sports car and 2.5PI saloon models.

Unfortunately the car was never as reliable, nor as specialized, as it should have been, for the dead hand of Leyland was weighing more and more heavily on the department. There was really very little money to spare to develop it, the car had to run with many well-proven but not race-sharp rallying parts and—frankly—it was not as fast as the Falcons or the Lotus Cortinas.

1967

There was a further contraction of 'works' effort in 1967, though Ray Henderson and Gordon Birtwistle continued to help wherever they could find the resources and (with greater difficulty) the money to do so.

The British Saloon Car Championship project, with the fuel-injected Triumph 2000, was dropped after a single year, but the team managed to continue the same arm's length support to Roy Fidler's rallying programme, when he used the same car (FHP 993C) throughout the season, though he did not have as much success. He managed fifth place in the Circuit of Ireland, when his partner was the rotund Attis Krauklis, a *Motoring News* journalist, and took sixth place on the Gulf London, which was rapidly developing into a British 'Marathon', even though he rolled his car on the first day, spent the rest of the rally with a crumpled car, and drove for many hours without a windscreen.

If all had gone well the climax of the season *might* have come on the RAC rally, where a new set of regulations allowed Group 6 prototype cars to compete. As the new TR5 (complete with its 2.5-litre six-cylinder engine) had just been revealed at that time, Ray Henderson—urged on by Harry Webster—went one stage further than he had been allowed to with the Bill Bradley race car, and produced a pair of what were effectively prototype 2.5PI rally cars!

The 2.5PI did not actually go on sale until the end of 1968, but for November 1967 Standard-Triumph produced two prototype rally cars. One, for Roy Fidler to drive, was an up-dated version of his own rally car (FHP 993C), but a second car (GVC 689D) was prepared for a promising driver called Denis Hulme to drive!

Hulme, in fact, had just won the Grand Prix Formula One World Championship in Brabham-Repco cars, when he was approached by the rally's sponsor, *The Sun* newspaper, to take part in the RAC rally. His racing rival Graham Hill was due to do the event in a 'works' Lotus-Cortina, which was an added reason for the New Zealander to take part in an end-of-season 'jolly'—the problem was that Denny had never before competed in a rally.

Denny finally agreed to tackle the event when *The Sun* made it financially worth his while, after which the paper asked me to take part as his co-driver. As I was then working for *Autocar*, that magazine was delighted to let me mix business with pleasure—and he was then guaranteed a day's pre-event testing and familiarization at the Bagshot test track in a virtually standard 2000.

That was a day when both Hill and Hulme were present, when Hill inverted his Lotus-Cortina practice car, and when Hulme discovered all about gravel track handling (we went off once, or twice, or three times . . . who knows?), but it was also a day which convinced him that he could finish well up, certainly in the top ten if the car did not let him down.

The prototype cars were more powerful and torquey, and at least as strong, as the old Weber-carburated 2000s had been in 1965, so everyone was looking forward to a good result. Because of Denny's Goodyear connections, we agreed to use Goodyear, rather than Dunlop, tyres on the event (Ford used the same rubber at this time, so they were expected to be competitive), and we were to use Minilite cast magnesium wheels. This, incidentally,

In 1967 The Sun newspaper arranged for the new Formula 1 World Champion, Denny Hulme, to drive a works Triumph 2.5PI prototype in the RAC rally, with the author as his co-driver. The event was cancelled just 12 hours before the start. Pictured at the Jaguar factory, where Denny had just picked up a new road car, are (right) Denny Hulme, and (left) the author.

was the very first time that anyone had ever *given* me a rally jacket—a Goodyear item!

However, even before scrutineering, an outbreak of Foot-and-Mouth disease in the countryside caused wholesale re-routings to be proposed. The cars went through scrutineering, Denny went home to Surrey, and I was actually working out a service schedule, when—on the evening before the start—the organizers cancelled the event.

Preparation of the prototype 2.5PI saloons was well advanced when Denny Hulme (centre), the author (left) and Ray Henderson (right) inspected the engine bay of one of the cars.

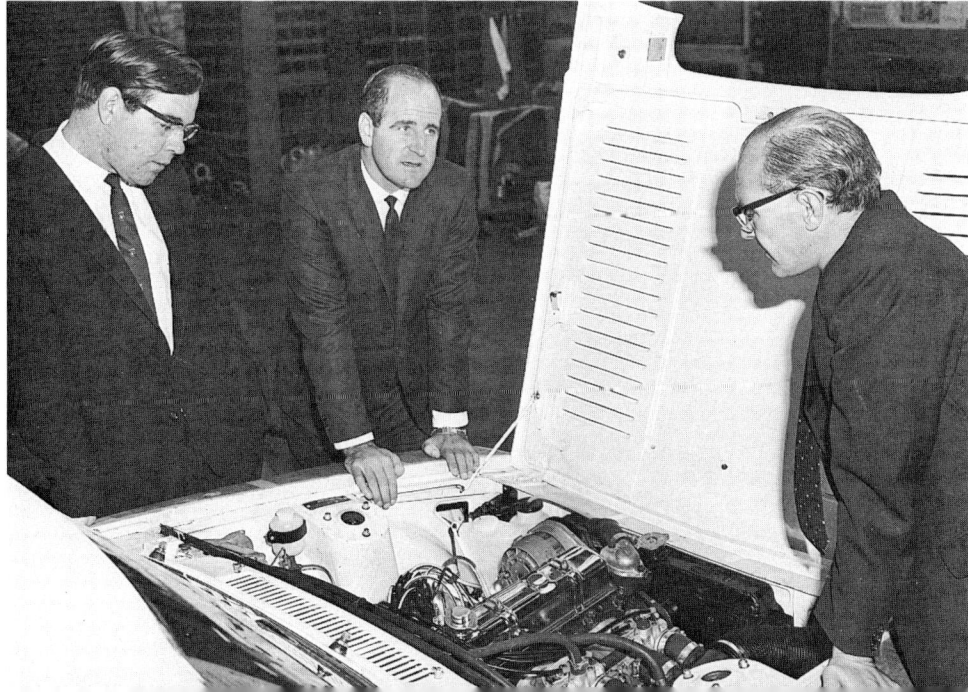

Denny Hulme in the process of learning that rallies don't take place on the same surfaces as F1 races! Testing, at Bagshot, before the 1967 RAC rally, which was cancelled just hours before the start. GKV 305D was merely a test 'hack', and was never used on any rally or race.

Naturally we were disappointed (though Denny later admitted to me that he had been very nervous about making an early mistake), but some people suffered even more than we did. Foreign teams had arrived, had changed all their money into sterling before the off, found the event cancelled on the Friday night, then heard Prime Minister Harold Wilson devalue the pound on the Saturday evening!

Soon afterwards, incidentally, I persuaded Harry Webster to lend GVC 689D to *Autocar* for its *Given the Works* road test treatment. Mike Scarlett, who wrote the words, noted that the final drive ratio was 4.55:1, which with a limited slip differential, and that there was more than 150 bhp available. He also stated that the car had standard gearbox ratios, even though it had been fitted with the rally-proven wide-ratio gears for the event itself, and that the rev-counter was red-blinded over 6,000 rpm.

Mike wrote:

> Like all the rally-prepared cars we have tried, the Triumph handled beautifully. Its straight-line stability was excellent, the ride reasonably flat and we quickly felt absolutely confident in it. On a track, lurid tail slides were easily provoked by chucking it into bends too early and fast, then holding the back out with power. . . It was only on the type of tracks that special stages are held that all the virtues of this glorious animal can be truly proved. The first thing that impressed us was the obvious strength of the car; the only limitation to how fast you take the jumps is your own resilience rather than the machine's.

It was interesting to see that this massive (2,900 lb.) rally car reached 100 mph in 24.5 sec., and rushed up to a rev-limited top speed of 117 mph, which was actually faster and more long-legged than Ford's rally-winning Lotus-Cortina could achieve at the time. If only the event had taken place . . . if only.

1968

The RAC rally 'non-event' preceded the run-down of the Coventry-based Triumph motorsport operation, for soon after this event Ray Henderson's team was dispersed and the two prototype 2.5PIs were eventually sold off. It was not, however, to be the end of Ray Henderson's work on competition cars, for in the next few years, when Triumph's motor-sport activities were based at the new 'British Leyland' Motorsport department at Abingdon, Ray was a necessary, and quite invaluable, link between Peter Browning's team, and the engineers in Coventry.

Ray Henderson was eventually obliged to turn to more mundane matters connected with engineering, but still found time to produce one final indulgence for Harry Webster and Spen King—the four-wheel-drive Triumph 1300 which Brian Culcheth used in rallycross in 1969. Gordon Birtwistle, who did all the test work on the car, remembers it as a glorious lash-up, the sort of car which made no engineering sense, but which worked—and worked well.

The technical story revolved around two inter-related projects—the 1300 saloon, and the lightweight 4 x 4 'Pony' (a small military-type vehicle built in Israel), which shared the same *basic* engine/transmission installations, but with different purposes. In the 1300 saloon the usual Triumph type of 1.3-litre Herald/Spitfire engine was mounted in the 'north-south' position, and was mated to a new front-wheel-drive installation, which featured an all-indirect four-speed synchromesh transmission mounted behind and below the engine, with the gearbox sprouting from the tail of the transmission casting.

This was carefully detailed to allow a propeller shaft to be hooked up to the output shaft, so that the rear wheels could also be driven, this being the basis of the 4 x 4 'Pony' installation.

At Harry Webster's request, Ray Henderson produced a special car, which mixed a Le Mans specification engine, the 4 x 4 'Pony' transmission, and the semi-trailing arm independent rear suspension of a Triumph 2000! In the beginning this was intended only as a cheap and cheerful car to be used in the London Motor Club's TV Autopoint event, an automotive orienteering 'party game' which Triumph had supported for some years. Such a car could not have been used in rallies, where four-wheel-drive was specifically banned.

Apart from its use of Minilite wheels, the only visual give-away on this car, which suggested that everything was not normal, was that there was a huge bulge and scoop in the bonnet to make space for the big dual-choke Weber carburettors

Gordon Birtwistle said:

> This car lent itself to having a Triumph 2000 rear suspension layout. It was just a case of putting in a prop-shaft tunnel, then grafting the 2000's cross-beam and arms under the tail.
>
> The engine, by the way, was actually a 1500, not a 1300, but built to Le Mans spec.—and it was possible to disconnect the rear-wheel-drive, by a lever under a flap at the side of the gear lever.
>
> I remember running this car on the road one weekend, and only using front-wheel-drive at first. It was an awful car in two-wheel-drive, and, by the way, it didn't have bumpers, and no silencer either. Because it didn't handle very well on tarmac with knobbly tyres, and I was getting harassed by boy racers, I slotted it into four-wheel-drive, and after that it just walked away!
>
> By the way, did you know that there was a mismatch on final drive ratios—and there wasn't a centre differential to even things out? We had a bit of difficulty with this at first—but it was a very nice mismatch for the driver to use, because it encouraged a bit of oversteer!

[After this comment I checked back into the archives, and saw that a normal 1300 or 1300TC would have had an overall final drive ratio (in front-wheel-drive) of 4 37:1. The normal Triumph 2000 rear-axle ratio was 4 1:1, though it would have been possible to machine a 4 3:1 ratio (as homologated for TR4s) on to the same forging blanks. That's a 1.7 per cent mismatch, but on knobbly tyres, and on loose surfaces, the transmission could cope with it!]

If British Leyland had not been founded in January 1968, this car might never have been used in serious events—but, as we shall see, it reappeared as an 'Abingdon-owned' machine a year later.

British Leyland—what happened after the merger?

After Leyland merged with British Motor Holdings (which included BMC *and* Jaguar) to found the British Leyland colossus, there was a period of great uncertainty throughout the Corporation, when all manner of rationalization plans were proposed, discussed, modified, and then enacted, often in a muddled and inefficient manner.

Right from the start, it seems, Lord Stokes (who was by no means a motor sport enthusiast) was determined to collect every competitions activity under one roof. Rightly or wrongly, he decided to concentrate British Leyland's efforts at Abingdon, where the famous BMC competitions team had been based since 1955, and where there was still a large and busy programme

Almost automatically, therefore, BMC's competition manager—Peter Browning— became British Leyland's motorsport supremo, which meant that Ray Henderson's moribund operation would have to report to it in due course

Peter Browning told me:

> My brief was to liaise with all the chief engineers. It was all very formal—there was a written brief, a Policy Instruction with a certain number. First of all, though, I have to remind you that things were never integrated in my time. You still had Triumph, you still had Rover, you still had Austin and Morris and MG—they were very competitive still. . . .
>
> Harry Webster moved over from Triumph to Austin-Morris at Longbridge. I was therefore dealing with Spen King at Triumph, Peter Wilks at Rover, and Harry Webster at Austin-Morris. The only two people who really had competitions experience were Ray Henderson, and Ralph Nash at Rover. . . .
>
> I never actually got central control. I got an instruction that we had to look at using other cars in the Group because we—BMC, that is—were not winning any more, the Mini was getting long in the tooth, and they were not prepared to invest in the development of fuel injection engines or bigger wheels.

Peter's brief, therefore, was to incorporate Triumph into any corporate programme which he got approval for—and that deserves a chapter all of its own.

Flying the flag in the USA

Strangely enough, while Triumph's British and European operation was winding down, 'Kas' Kastner's North American activities were going strongly. By this time 'Kas' was well-respected within Standard-Triumph. In 1965, indeed, it had even been suggested that he might move to Europe to take over the office which I had just vacated, though this was never pushed very far.

Before and during every season, 'Kas' played the SCCA game as craftily as only he knew how. By 1967, when the GT6 had gone on sale, he made sure that three different Triumph sports cars—Spitfire, GT6 and TR4A—were *all* competitive, and that they never had to compete against each other:

> I expect there was someone on the SCCA board who knew what I was up to, and was as bright as I was, but it wasn't evident. I like to think I played the cards pretty close to my chest.

Although Bob Tullius was the most famous and successful driver, 'Kas' never actually operated a 'works' team from his Gardena premises, where he concentrated on research and development.

When the TR250 (the USA version of the TR5) went on sale in the USA for 1968, the SCCA board hit Triumph hard, by listing the car in Class C instead of Class D where the TR4/TR4A cars had been so successful:

> The SCCA were totally afraid of six-cylinder cars. In any case, for our purposes, the TR250 had a terrible engine at first.

One of Triumph-USA's most successful drivers, Carl Swanson (left), and 'Kas' Kastner discuss progress already made on an early racing example of the GT6. This shot dates from 1967, the first year that type of car was raced.

It was hard to work on. Harry Webster defied me to turn it into a race-car engine—he said the crankshaft would come apart at 6,500 rpm. But what hurt us most was that we were not allowed to run fuel injection, even though the European TR5 had it as standard. The TR250 road car had no more power than the TR4A, but we had to compete in Class C with much faster cars.

Bob Tullius, by far the most successful Triumph driver in the USA in the 1960s, with an early example of the Group 44 GT6 cars with which he regularly won his SCCA group—in this case E Production (hence 'EP').

Turning the six-cylinder engine into a rugged and reliable racer was a saga which 'Kas' never wanted to repeat, but before he had finished, the 2-litre version (as used in the GT6) was producing 218 bhp on Weber carburettors:

> The 2-litre was a better engine than the 2½-litre. There wasn't any overlap on the bearing surfaces of the crankshaft of the 2½-litre—really, it was a piece of rubber. The third vibration period peak was so steep that it went off the page on the recorder—and that was at 6,200 rpm.

Remember that 'Kas' had no help from Coventry (for he knew far more about race-tuning that the factory ever did) and you realize the magnitude of what was achieved. First he had to make the flywheel stay attached to the crankshaft, then he had to keep the timing chain in place, and in the end, after commissioning a vibration analysis, he actually *added* weight to the crankshaft by bolting on extra weights:

> Eventually it was OK, not only could we get through 6,200 rpm, but I could go all the way to 7,200 rpm. I had my own air-flow bench, so I could get the cylinder head to breathe—especially as the factory produced special heads from 'scrubbed' cores, so that I could mill more than 200 thou' off the face to get the compression ratio up.
>
> In Coventry, well I had to charm the engineers into helping me. I couldn't have a five-speed gearbox when I needed it—to match Datsun—but George Jones educated me about gear design, and showed me blank forgings. Then I asked Harry Webster to machine up some special ratios for me, and I got them homologated with the SCCA.

When I asked 'Kas' to summarize his eight years with Standard-Triumph in the USA, he told me that the TR4 was the easiest model to turn into a successful race car ('but that was because I had so much experience with my own TR3A'), that the TR250/TR6 was the most difficult because of the engine, and the GT6 because of the rear suspension:

> We'd worked out how to make the Spitfire handle, by using a camber compensator, but the GT6 had a lot more power—even with twin Strombergs we were running 185 bhp, with the engine turning over at 8,200 rpm. It was a great engine, the 2-litre, and we ended up putting the 2½-litre cylinder head on it, which was a good move.

The K-Car

The most exciting project Kastner ever tackled was abandoned soon after birth—in fact it seems to have died of neglect. This was the TR250K—the 'K-Car'—of which just one example was built. Kastner is still puzzled, if not bitter, about its fate:

> It all started because of my friendship with Pete Brock [in motor racing Brock was already known for his styling work on the ground-shaking AC Daytona Cobras of 1964]. Pete and I wanted to do a design on a production car. Pete did some sketches on the basis of the TR, I showed them to Chris Andrews, and he said we should show what we could do, and maybe make good advertising out of it.
>
> I asked Pete for a proper sketch, so he spent about an hour producing an 18 in. drawing on poster paper. I was on the way to England, but went by way of New York, and saw Leon Mandel, an old friend of mine who was Editor of Car & Driver. I put a deal to him, saying:
>
> 'We're going to make this car. Will you put it on the cover of your magazine?'
>
> Well, he agreed to that, I went on to see George Turnbull, who was running Standard-Triumph for Stokes by then, and said I was sure of exposure on a couple of magazines, and maybe several more stories, and said we were looking at maybe $50,000 to $100,000 of publicity.
>
> Turnbull then said: 'OK, go ahead and build it'—and said he would pay for it, at a cost of about £20,000. I gave Pete the go-ahead. He handled the building of the body which was all aluminium, and I did the race-preparation of the chassis, which included wheels from one of the Chaparral cars—and that was the part that failed.

The beautiful Triumph 'K-Car', as prepared to take part in the Sebring 12-Hour race of 1968. The chassis and running gear was 'Kastner-Triumph' of the period, complete with a six-cylinder engine, but the body was a completely new style, by Pete Brock, who had also shaped the Daytona Cobras of 1964. Pete Brock is on the left of this group, 'Kas' Kastner is third from left, and one of the drivers, Jim Dittemore, is on Kas's left.

Officially called the TR250K, but more familiarly as the 'K-Car'—where 'K' naturally stood for Kastner—the Brock-styled project was a magnificently shaped machine, with a long low snout, smooth flanks, a high and short tail, and looked several years ahead of its time; though it was never tested in a wind-tunnel:

> Donald Healey saw it at Sebring, and told me it was the most beautiful thing he'd ever seen. We even had a movable spoiler at the back, a section about five inches wide, the full width of the car. There was a lever inside the car connected to the flap by a Bowden cable. You pulled on that and the flap came up. It could be adjusted to whatever notch you wanted. I think that *may* have been one of the very first race cars [after the Mercedes-Benz 300SLR] to have a movable aerodynamic device.

Another view of the TR250K, or 'K-Car', which 'Kas' Kastner of Triumph-USA built to compete in the Sebring 12 Hour race of 1968. Jim Dittemore (at the wheel) and Bob Tullius were the two drivers. ('Kas' Kastner)

By 1968, when this Weber'd 2.5-litre engine was fitted to the TR250 'K-Car', Triumph-USA was able to produce a lot more power than any Triumph-UK built engine had ever managed to do. ('Kas' Kastner)

Kastner admired it then, and still admires it today:

> I've never been in love with another piece of machinery since I had to sell my first MG, but this was a great shape. It was a convertible you could drive without your hair being tugged out—it was pure 1980s Corvette.

But this was 1967/1968, and Kastner entered it for the Sebring 12-Hour race in March 1968. It was raced as a Triumph prototype, with Kastner's favourite drivers, Bob Tullius and

The first rallycross-winning Triumph was this extremely special four-wheel-drive 1300, driven in the 1969 winter series by Brian Culcheth. (Brian Culcheth)

Somewhere behind all that mud is Brian Culcheth, driving the four-wheel-drive Triumph 1300 rallycross car, which won several races, but was later written off when the 2000-derived rear suspension broke. Note that the rear doors of this car have been welded closed. The big bonnet bulge was necessary to make space for the Weber carburettors of the Le Mans-tune of engine. (Brian Culcheth)

Jim Dittemore, at the wheel. Unhappily, it ran for no more than 90 minutes before the right rear wheel failed.

What happened next?

> Nothing. I brought it back to California, where it just sat. No-one came from Triumph to see it. No one seemed to be interested. It went to one of the car shows, but that was all. We'd originally hoped that someone—like Harry Webster, or Bruce McWilliams from the importers in New Jersey—would take an interest.
>
> But there was nothing; it was just as though it had never existed. I think it was a case of Not-Invented-Here-ism—there was certainly an element of jealousy involved. Eventually I ended up in possession of it, to compensate me for the time I'd put in. Then I traded it to Pete for the Samurai which he'd built, then ultimately I sold the Samurai. Both of these cars are now in museums—the K-Car in the Black Hawk museum, in Northern California.

At about the time that the K-Car was frozen out, this was perhaps the time that Kastner's love for Triumph began to fade. He'd tried very hard with the K-Car, but been spurned, and I am sure he was hurt that his genius at extracting power from engines was never used in other 'works' cars. He was rarely even consulted. Yet:

> I always got on well with Lord Stokes. He was very supportive. I thought he was a great guy. In my mind he didn't do anything with British Leyland that I wouldn't have done in the same position. When he visited me in Gardena I once put him into a GT6 with a 2½-litre engine. That was the trip when I also took him to a topless bar.

It wasn't only 'Kas', though, who had found out that life under British Leyland control was going to be difficult. Back in England, the 'works' operation was on the move.

Chapter 7

Marathons and Safaris
A change of direction, 1969-1973

From the beginning of 1969, British Leyland's policy of centralization meant that almost all future 'works' Triumphs were to be prepared and entered by the Competitions Department at Abingdon, though Ray Henderson carried on working on competition cars in Coventry. Politically, and practically, it was a difficult situation for both Peter Browning and Ray Henderson to accept, and—as Gordon Birtwistle has reminded me: 'there was a great deal of Them and Us at the time.'

The whole emphasis of British Leyland—and Triumph—in motorsport changed between 1968 and 1972. It wasn't that there was a strategy, or even a short-term policy—it merely seemed to be expediency. Mainstream European rally programmes were killed off in favour of unrelated entries where particular cars might just have a chance of success. There was one huge effort, however, which involved 2.5PIs being built to take part in the *Daily Mirror* World Cup rally, to Mexico City.

By 1972 Ray Henderson had built the prototype Dolomite Sprint rally car, and testing had begun, but there was no chance of using it until it was homologated for 1974. In a few short years, then Triumph achieved one of its best results, then dropped out of the scene almost completely.

Peter Browning was instructed to shop around in the corporation, to meet all the engineers and to see 'what was in the cupboard'—to find out what was coming in the future—then to evaluate a programme of possible motorsport activity for the future. But Peter had an uphill task:

The problem was that British Leyland was only prepared to give approval on a very short-term basis, there was no long-term commitment. In my view, you just shouldn't have an unrelated go at rallycross, or saloon car racing, or rallying, or whatever. I wanted to decide on a route, and follow it, to go on as long as it looked promising—but to *expect some failures* along the way. No-one seemed to understand this.

Everything we wanted to do to the cars had to be approved—all the technical specifications, and all the modifications, had to be approved by chief engineers. You can just imagine the problems—you simply don't run a Competitions Department on that basis.

I had to deal with people who, frankly, were not inspired by competitions. These were guys who would get very excited about competitions when the cars were successful, but who didn't want to know when we lost. They were very intolerant of failure. They could not understand that this was a game that you could sometimes lose.

Events like Le Mans though—well, they were important! Spen King once told me that the entertaining bill at Le Mans in the 1960s exceeded the budget for preparing the cars.

But Lord Stokes was never enthusiastic about us. There was great jealousy of MG, both from the sales and the competitions point of view. George Turnbull, however, of all the people I dealt with, was superb. He was very fair, very understanding, and quite enthusiastic. Incidentally, I *never* came face-to-face with Lord Stokes, other than when he decided to close the Department down. At the time it was George Turnbull who had executive responsibility; he approved all the budgets.

Ray Henderson was always very good, there were no chips on his shoulders, and I always found Spen King to be very courteous, very friendly and very helpful. He'd done some competitions in his time, and he seemed to be very enthusiastic. However, I know he had an awful lot of production pressures at the time, which meant that he couldn't give a great deal of time to our particular needs.

Then there were all the usual problems in starting afresh. In 1969, the department at Abingdon had no spares ('not a nut and bolt in the place') for either Rover or Triumph cars, and everything that was needed had to be requisitioned, and took ages to arrive.

Our other problem was that Neville Challis seemed to have friends all through BMC who could help us, and could make things quickly when we needed them. But with Triumph we didn't have that sort of relationship. First we had to go to Spen King, then to Ray Henderson—but Ray didn't have a budget. At first, when we said we wanted four diffs, people would come back with: 'What do you need *four* diffs for?'

It was almost like working for a foreign manufacturer. It was a clash of philosophies—they didn't understand our way of doing things, and in any case they didn't have budgets for that sort of thing.

Even so, Browning obediently jumped through all the organizational hoops and came to a conclusion:

We dismissed other projects pretty quickly. Apart from the Range Rover, and getting a hot Traco engine out of Ralph Nash at Rovers for the racing project, we considered the 3500S for rallying. I felt that the Rover *could* be used for rallying, on the basis that there was no substitute for horsepower.

Other Triumphs? Well, we were shown various new projects by Spen King including the Stag, God help us, which wasn't suitable. The Dolomite was still under cover—there was certainly a future there, but not yet. . . .

Brian Culcheth was used to assess each car—he was a more committed test driver than Paddy Hopkirk, especially as he had plenty of time to do the job, whereas Paddy was already building up his businesses. Browning agrees with this:

He didn't contribute a great deal to the testing, to development. He didn't like doing it, it was below his dignity a bit. I don't think he was good at translating problems to engineers. Brian Culcheth was more meticulous—he was very patient. The mechanics just loved him: Culch never let us down.

Paddy, however, as an ex-Monte Carlo rally winner, was a superb PR man. He was very good at that. We used him a lot, sometimes to bridge the gap which I couldn't bridge, in relationships with the Top Brass at Triumph.

It soon became clear that Triumph's new fuel-injected 2.5PI saloon—the descendant of the 2000—was the most promising of the bunch, and this was chosen for development. For Triumph, and for Abingdon, it was to be a busy and fascinating period—but it was also very short. The 'official' works 2.5PI period covered less than two years, though Brian Culcheth carried on the good work for another two years after that.

1969

Abingdon's very first link with Triumph came in the development of the 1300 4 x 4 saloon for use in rallycross. Soon after the car had been used in the London Motor Club's TV Autopoint event:

'Ray Henderson rang me,' Brian Culcheth says, 'and said they didn't want any idiot to drive it, so would I be the chosen idiot. . .'

It was delivered to Abingdon, where Peter Browning and Bill Price were interested in using it in TV rallycross events.

Peter Browning makes no bones about his reasoning:

> I was looking for instant action, not big-budget expenditure. Hopefully, I also wanted televised success. I hoped this would reflect well on us, and show management what we could do.

As Bill Price recalled in his book *The BMC/BL Competitions Department* (Haynes Publishing):

> It was very standard looking when it arrived at Abingdon, but now had the steel removable panels replaced with aluminium, and the inner door mechanisms stripped out along with the rear seats, trim, bumpers, etc. Perspex windows were also fitted to the side and rear to reduce weight, the only obvious exterior change was the blanked off headlamps.

In fact the rear doors were also welded closed, and the exterior handles were removed.

To gain the maximum publicity which Lord Stokes was demanding, Peter Browning's team was concentrating more and more on circuit racing and rallycross. Since the 1300 4 x 4 was not a homologated machine, and since four-wheel-drive was thought to be promising for rallycross, it was decided to use it in the TV Winter Rallycross series.

To drive it, Browning hired Brian Culcheth, the young man who had once driven a 'works' TR4 in the Spa-Sofia-Liege event, and who had eventually joined BMC to wrestle with cars like the BMC 1800 'Super Landcrab'. Once the 'Super Finns'—Timo Makinen and Rauno Aaltonen—had left the team, and Tony Fall had been attracted away to try his hand with Lancias, this left Paddy Hopkirk and Brian Culcheth as the regular team members, and to evaluate the various cars.

Incidentally, to shatter the myth that professional drivers made a lot of money in those days, Culcheth recalls that his retainer fee was £2,700 for the season, and that Paddy Hopkirk earned more, but not a lot more, as the 'star' of the team.

The 1300 4 x 4 had a short and spectacular career. Starting at Lydden Hill, in snow and ice which made it ideal for four-wheel-drive, the car annihilated everything else on live TV, which pleased everyone. Culcheth was delighted ('It took two seconds off everybody in the first 300 yards, and the handling was excellent') and Browning was relieved: 'We couldn't have got off to a better start. On the Monday morning the phones never stopped.'

The following month the car appeared at Croft and won a demonstration race, but didn't figure at the same circuit a month later after the gear lever broke when the car was in the lead. Then, in April, came the breakage at High Eggborough and the crash, where a rear suspension arm broke, pitching it into a dramatic roll, and writing it off.

There's a picture of the car actually 20 feet in the air, with the trailing link already broken, and with the car well on its way to destruction. In Browning's own words: 'That was the end of *that* project.'

In the meantime, Browning had finally decided to concentrate his rallying efforts on the Triumph 2.5PI, which meant that his team of mechanics, at Abingdon, would have to learn all about the car, from the ground up. Testing at Bagshot proved that the hull, and the chassis in general was more rugged than any other car in Abingdon's recent history. For rallying, though:

> The first car we used was a Mark I, which had been totally prepared by Ray Henderson's people in Coventry. It came down to Abingdon, we went over it, ordered a few bits and pieces as spares, then we sent it off, more or less as a testing rally, on the Austrian Alpine rally, for Paddy Hopkirk to drive. I went along with someone from Triumph, and did the servicing with two of our mechanics. I'm afraid it retired with clutch trouble.

A new car—which nevertheless carried the same identity (UJB 643G)—was then prepared at Abingdon for Brian Culcheth to use in Group 6 (prototype) form in the Scottish rally. A new car with an existing identity? Don't be shocked—that sort of thing has been going in 'works' motorsport departments since time immemorial. The gaff was finally blown in November when *both* cars were used at a pre-RAC rally test day!

In this tune the 2.5PI had Minilite wheels, was running without bumpers and had some aluminium panels. Compared with the BMC 1800s he had recently driven in rallies, Brian found the PI very much faster, and held down a strong third place overall for some time. Then the rear axle failed, and as John Davenport's *Autosport* report pointed out:

> The biggest story of the day was that of Brian Culcheth, who broke a differential. When the BLMC mechanics came to drop in the new unit, they found that its front brackets had been welded to the trailing arm carriers, and stubbornly refused to become unwelded, despite much coaxing from Den Green and his men. The whole job ran to two hours which meant that Culcheth and Syer had to miss two stages—at 1,800 marks a time—to get to Aviemore on time.

The result was that the car eventually trailed in to finish 24th overall, and second in its class.

For the next few months there was no evidence of 'works' Triumphs in motorsport, while British Leyland concentrated on racing its Mini-Cooper S cars, but by the end of the summer a series of rumours concerning a marathon rally in 1970 had been confirmed. The *Daily Mirror* had agreed to sponsor a 16,000 mile trek from London to Mexico City—by way of Sofia, Lisbon, a ship to Rio de Janeiro, then a complete clockwise circuit of South America.

This ambitious event was scheduled to start from Wembley stadium on 19 April 1970, and to reach Mexico City on 27 May 1970. Because it was to end in Mexico City, just before the World Cup football tournament took place, its title was obvious—this was to be the *Daily Mirror* World Cup Rally.

More than 20 years on, every rally historian is agreed, that this was the toughest rally of all time, especially as three factories—Ford, Citroen and British Leyland—all took it very seriously indeed. Almost as soon as he heard of the event (for which ex-Triumph team driver John Sprinzel was to be the organizing genius), Peter Browning went to his bosses, con-

Brian Culcheth and Johnstone Syer lost a high placing on the Scottish rally of 1969 when their Abingdon-built 2.5PI suffered rear axle problems, which could not be replaced in time to keep them in the event. In specification, this car was a direct descendant of the cars developed at Coventry in the mid-1960s. (Brian Culcheth)

The 'works' team posing by three newly-built 2.5PI 'Mk I' cars, ready for the RAC rally of November 1969. Left to right: Brian Coyle, Andrew Cowan, Paddy Hopkirk, Tony Nash, Johnstone Syer and Brian Culcheth. (Gordon Birtwistle)

vinced them of the benefits of successful competition, and secured a big budget for the event. Andrew Cowan, who had won the 1968 London-Sydney Marathon in a Hillman Hunter, joined the team for this event.

The complication was that the existing 2000/2.5PI range was due to be re-styled, though the general layout and chassis was not to be altered. The new cars would appear at the London Motor Show in October 1969. Accordingly, the master plan (and for such a vast operation it *had* to be a master plan) envisaged the use of Mark I shape cars on the RAC rally, then using them as test, development and recce cars for the World Cup Rally, but to

Paddy Hopkirk's 1969 RAC rally 2.5PI was one of the last three 'Mk I' shape cars to be prepared for 'works' motorsport. In Group 2 form it had to run with bumpers, but was fitted with Minilite wheels. (Gordon Birtwistle)

build a set of new Mk II-shape cars for the rally itself:

> We had a very good budget for the event, and Lord Stokes was excited about it. We had to say—
> and we did say—that we could win. We wouldn't have been allowed to go at all if we were not
> sure to be front runners. They (the management) were convinced that we were going to win for
> them, but we were always worried about the big Citroens, and about Ford's Escorts.
>
> From late 1969 until the summer of 1970, the World Cup Rally took up *all* of our resources.
> Incidentally, I've forgotten what the budget actually was—and I'm not being evasive. Probably
> because it snowballed, anyway—every time Culch came home again from South America he'd say
> that we needed to have a bigger aeroplane, more service, more wheels, more tyres, or something.

As a prelude, and to get more competitive experience with the 2.5PI, Abingdon then built
up three brand-new cars to use in the RAC rally of 1969. Even at this stage British Leyland
admitted that it looked on the event as one long test and learning session for the World
Cup Rally—and explained that it had to use Mark I shape cars because Mk II types could
not be homologated in time.

In effect, therefore, these cars were instantly obsolete, and were entered in Group 2 form,
their fuel-injected engines having about 140 bhp, and the final drive ratio (as on the
Scottish rally) was 4.55:1; the 13 in. wheels were alloys by Minilite, with 7.0 in. rims.
When pressed for a forecast, Peter Browning told the media that he thought his cars would
finish third, fifth and ninth.

On a 'normal' RAC, where the tracks were loose, but where there was little snow, the
2.5PIs should have been excellent machines—but this was not a normal year, for there was
a great deal of snow and ice on many stages. Ironically enough, British Leyland would probably have won the event if Abingdon had been allowed to use front-wheel-drive Mini-
Cooper S types, for the event was dominated by Lancia Fulvia HFs and Saab V4s, both of
which *had* front-wheel-drive.

Because of the weather conditions, it was all too much for the big and heavy Triumphs.
Paddy Hopkirk resorted to chains at one point to keep his car moving, and all went off from

*Paddy Hopkirk and Tony Nash on the 1969 RAC rally, in their Abingdon-built 2.5PI on an event
with appalling weather. When asked by a TV interviewer: 'What is the most difficult thing about
this rally?' the irrepressible Paddy retorted: 'Winning it in this motor car!' (Gordon Birtwistle)*

Brian Culcheth rapidly came to terms with the big Triumph 2.5PI, enjoying its solid strength, and its supple suspension. Here, on a very icy stage of the 1969 RAC rally, he is on his way to third in class. (Gordon Birtwistle)

time to time. Even so Andrew Cowan, who started in the No. 1 slot, was 11th overall at the half-way point.

As it turned out, the biggest battle for the 2.5PIs was even to get to the finish, and for the Manufacturers' Team Prize. At the end of an event which had included no fewer than 73 special stages, all three cars finished—first, second and third in class, with Andrew Cowan

Andrew Cowan, who won his class and finished 11th overall in the RAC rally of 1969, must have thought he was on the Monte at times, especially when the snow was so deep in Greystoke forest, in the Lake District. (Hugh Bishop)

still in 11th place overall. In the Team Prize contest the gap between Triumph and Datsun was seconds only—but Datsun took the award.

Paddy Hopkirk had every reason to be disgruntled by the event and the problems which he had had, but was there *really* any reason for him to do his 'irrepressible Irishman' act in front of the TV cameras?

'Tell me, Paddy,' said the *Wheelbase* interviewer, 'what is the hardest thing about this rally?' To which Paddy replied, with a grin: 'Winning it in this motorcar!'

The World Cup saga

In some ways, preparations for the World Cup rally began in the summer of 1969, when British Leyland very craftily arranged to lend the ex-Austrian Alpine/ex-test 2.5PI to John Sprinzel and John Brown to carry out their route survey of the South American sections. British Leyland's own surveys began immediately after the RAC rally, when Brian Culcheth and Tony Nash flew off separately to South America to begin a recce in a Peruvian-registered 2000 which Uldarico ('Lacco') Ossio had driven across the continent. For the next five months Culcheth almost became a South American resident:

> I was the first to go out to South America, to Brazil, to start our first recces. Tony Nash joined me a week later, in Santiago. Really, I did almost three complete World Cups on those recces.
>
> Due to what we found out in that time, we came to a decision that these recces were too rough for standard cars, or hire cars, so we decided to ship out the three ex-RAC rally cars. The whole team—Paddy Hopkirk, Andrew Cowan and myself—started proper reconnaissance after Christmas 1969 in those cars.
>
> On the other hand we didn't bother much with the European sections—we had notes for most of them from other rallies. I had six months in South America, but just six weeks in Europe!

Peter Browning told me that at this time the department thought only of the World Cup rally for months:

British Leyland put an enormous effort behind its entry for the Daily Mirror World Cup Rally of 1970. Paddy Hopkirk (right) and Brian Culcheth (centre) discuss their 'Survey' (recce) car with two of Abingdon's mechanics.

The Triumphs were the spearhead of the team. But don't forget we built two Maxis as well, for Rosemary Smith and the Red Arrows to drive, and a Mini Clubman for John Handley. We used Maxis because Range Rovers were not available.

Should the 2.5PIs go two-up, or three-up? This was a big decision to make, and a controversial one:

Culch came back from South America and said 'Definitely two', but the old men—Hopkirk and Cowan—wanted an easier ride, so they decided to go three-up.

Culcheth never had any doubts:

Firstly I thought it would be a lot of extra weight, particularly on the rear suspension, and secondly I wasn't convinced that having three people in a car was a good thing, particularly because of the strain that comes on such a rally. The sheer tension of having three different people, three opinions—I'd experienced this on London-Sydney (with Tony Fall and Mike Wood) so I felt that the two of us (Johnstone and myself) could handle it.

There was one critical section, from Santiago in Chile to Bolivia, which was 57 hours non-stop with three Primes [special stages]—that was the only one I had doubts about. But we felt that if we could get to the Bolivian border with enough time in hand, we could have three to four hours rest before we started the third prime to La Paz. As things turned out, it worked.

The tiredest I was on the whole event was on that last mile into service after that long Argentinian Prime. But we had an hour's rest immediately at the end of the stage. I asked the mechanics to wake me, then we pressed on to the Bolivian border, driving hour on/hour off. We got there and had four-and-a-half hours rest. That was the turning point.

By this time, in any case, Culcheth and Syer, like Andrew Cowan and Brian Coyle, were regular partners and firm friends, which made job-sharing very easy and straightforward:

Johnstone was a very good driver, but he had no ambitions as a driver. He got tremendous satisfaction from being driven fast on stages by good drivers.

Building the cars was a combination of Ray Henderson's expertise, the application of a mountain of experience by Abingdon's mechanics, and specialized help from suppliers.

In 1970 the 'World Cup' Triumph 2.5PIs used a completely new, and very complex, electrical wiring installation. There was a lot more work to be done before this car would be finished. (Brian Culcheth)

By the time British Leyland's Abingdon competitions department had re-equipped a 2.5PI for rally-ing, little of the glossy equipment of the standard car remained. This was WRX 902H, as built for World Cup rally test work in 1970. Note the overdrive switch on the gear-lever knob. (Brian Culcheth)

Pressed Steel, in particular, provided very special body shells, complete with foam-filled sills, many aluminium panels, and a great deal of extra strengthening, along with flared wheel arches to allow 15 in. wheels and tyres to be fitted, and with bulges in the boot lid to allow two spare wheels to be carried. There were cold air scoops in the roof, perspex side and rear windows, and the co-driver's seat could be converted to a bed.

Mechanically the engines were tuned with 16,000 miles in mind (the fuel injection included an adjustable calibration for the huge height variations on the event) with about 150 bhp. There were special gearboxes, limited slip differentials and three fuel tanks totalling 32 gallons.

Four cars were entered—the fourth being for the Australians Evan Green and 'Gelignite' Jack Murray, with Hamish Cardno (of Britain's *Motor* magazine) as 'third man'—and there were no team orders. Browning remembers that Lord Stokes expected him to win with one of the cars, and didn't mind which one it was. To cover every eventuality the parts were transported around the two continents in a four turbo-prop Britannia aircraft. Naturally this was immediately nicknamed 'Browning's bomber':

Peter Browning recalls:

> Ford service on the ground in South America was a hundred times better than ours, in terms of local dealerships. For that reason we chose to take an aeroplane, and to take parts from country to country.

The 'World Cup' was a phenomenal event, where almost every statistic was a superlative. Not only did it occupy six weeks and 16,000 miles, but it passed through no fewer than 25 countries, and had attracted crews from 22 countries. The most difficult Prime (special stage) covered no less than 480 miles of flat-out motoring, for which the unattainable target time was eight hours. Of that stage, Culcheth said at the time:

> It was rather like driving from Edinburgh to Dover, mainly in fog, on unmade roads strewn with rocks and animals, as fast as you could make it.

That was the Prime which also included passage of the 15,650 feet Agua Negra pass, while the even longer (560 miles in a target time of 11 hours!) Peruvian 'Incas' Prime included a

Before the start of the Daily Mirror World Cup rally at Wembley Stadium, British Leyland's chairman, Lord Stokes, chats to (left to right): Johnstone Syer, Brian Culcheth and competitions manager Peter Browning. (Brian Culcheth)

peak of no less than 15,870 feet. It was no wonder that all the serious crews arranged to carry oxygen for these sections.

Surprisingly, the first prize was 'only' £10,000. As Peter Browning made clear, this would only make a dent in the budget he needed to tackle the event, and he couldn't guarantee to win it.

By comparison with what was to follow, the London–Vienna–Sofia–Monza–Lisbon sec-

The Evan Green/Jack Murray/Hamish Cardno 2.5PI eases its way off the ramp at the start of the Daily Mirror World Cup rally, at Wembley Stadium.

Two grubby World Cup 2.5PIs being serviced at Monza, before going into parc ferme for the overnight halt. The cars had to have their door-positioned competition numbers covered up for the passage through Italy.

tion, which was mainly on fast main roads and which included only one overnight halt (at Monza) was a doddle. Two Primes in Yugoslavia, one near San Remo, one behind Nice (the famous ex-Alpine rally section of Quartre Chemins–Entrevaux–Sigale–Rouaine route, for which 90 minutes was allowed) and one in Portugal, merely whetted the appetite for what was to follow.

At Lisbon, one week after the start, three of the Triumphs were sixth (Culcheth),

Brian Culcheth and Johnstone Syer were still settling down to their 16,000 mile task on the World Cup Rally when they passed Castle Schoenburg, near the Vienna control in 1970.

eighth (Hopkirk) and eleventh (Cowan)—but Green's car was way down in 69th place, damaged and in serious trouble. Whereas the 'regulars' had driven carefully, and were about 30 minutes behind the flying Fords and Citroens, Green's car was already in serious engine trouble, and running on only five cylinders due to a valve guide being damaged, while in France it had also plunged off the road on the Alpine Prime when a front wheel had come adrift.

Even though the car was eventually retrieved, and the cylinder head changed at Lisbon, it was definitely 'walking wounded' (with the fuel from the appropriate fuel injector piped out through the wing vent), for it was more than 1,000 *minutes* behind the leaders.

For the next 12 days, the 71 cars (out of 98 which started) were carried carefully across the South Atlantic by the *Derwent*, which left time for the crews either to rest (many did), to carry out last-minute recces in South America or, as in the case of the author, who was operating as a travelling controller, to go back to work for a time!

The major part of the event, which took in many of the capitals of South America, now showed that this was no normal rally. It occupied three weeks, 12 more Primes (all long, but some of them quite colossally long), and every possible variation of terrain, temperature, and road surface.

This was where reputations would be made and lost, and where British Leyland's philosophy of building the 2.5PIs as 'fast tanks' would be tested. The route went from Rio de Janiero to Montevideo, by ferry to Buenos Aires, then to Santiago, La Paz, Lima, Quito, Buenaventura, another ferry to Panama, and a final burst up the Central American isthmus to Mexico City itself. At least half the nights were spent in the cars, and the target average speeds were always high.

At the end of the event, two of the big Triumphs would figure strongly in the results, but Brian Culcheth, who had also staked his reputation and experience on his lengthy surveys, recalls that:

> Basically, we always went as fast as we thought the car could stand. In Europe we had driven very gently, to keep in the frame. Then in South America two Primes were changed, which turned what should have been our advantage into a disadvantage, these were in Uruguay and Argentina. In Argentina the time schedule was easy, compared with a difficult recce, because

The Culcheth/Syer 2.5PI, on its way to second overall in the Daily Mirror World Cup rally of 1970. With only an extra bit of luck Culcheth could have won this event—at least his team didn't cheat by changing forbidden components along the way. (Brian Culcheth)

the authorities had regraded all the routes—they were like a motorway.

[In fact all the leaders cleaned both the Argentine Primes]

In Uruguay it was similar, Andrew Cowan and I were first and second fastest—Andrew by a minute, but I'd had to stop to change two punctures—and we beat the Fords by several minutes. I seem to recall that we *averaged* 108 mph on that Prime.

In Argentina I reckon we could have been an hour faster than the Fords if we'd all been penalized, because they had a very limited top speed, whereas we could do 125 mph. We didn't really have a lot of choice about gearing, because the 4.55:1 diff. was the lowest we could use without getting into crown wheel problems.

By mid-May, at Santiago, the capital of Chile, the event was still wide open. Rene Trautmann's Citroen had crashed and the other two Citroens were falling behind, while the Ford Escorts seemed to be in serious mechanical trouble. Although Culcheth's 2.5PI was 31 minutes behind Hannu Mikkola's Ford Escort, it was now up into fourth place, with Hopkirk's car seventh and Cowan's car tenth. The Green-Murray-Cardno car had finally blown its engine in a very big way, when a rod went through the side.

I was thumbing through memories of this event with Culcheth, when Brian suddenly burst out with:

By about Santiago, though, the Fords had all sorts of axle problems, and with Mikkola's reputation *then* (he hadn't finished a rally for about a year), we thought there was no way he could keep going to the end.

A year or so after the rally, at a party, one of Ford's mechanics said to me that: 'You were really the winners', because they had apparently changed Hannu's gearbox for one from another car, to keep it going, then changed it back when it was repaired. On that event you weren't allowed to change gearboxes, axles or the cylinder block, because they were all sealed and marked. If there'd been a check—there wasn't—he'd have been disqualified.

For the rest of the event, where the route wound its way up the West Coast of South America, and up to Mexico City, the 2.5PIs gradually pulled up the field. At the rest halt in La Paz—where the airport and the *parc fermé* were at an altitude of 13,000 feet—Culcheth's car had taken third place behind two Fords (65 minutes adrift), with Paddy Hopkirk sixth (his car had broken its back axle at an inopportune moment)—but only 39 cars were running, and Andrew Cowan's car had crashed badly, injuring the crew.

Cowan was tackling the 480-mile Gran Prime and making good time until, 27 miles from the end, he crashed when trying to pass another car in a thick dust cloud. The car was a write-off, with its roof line almost flattened down to scuttle level, and all the crew were badly knocked about in the crash, and had to be taken to hospital in Salta.

As a travelling controller on the event, I vividly remember having to visit the unhappy crew in hospital, to find Ossio suffering a broken skull (he had not been belted in) while Cowan suffered a cracked neck bone, and Coyle a broken wrist. Within days, fortunately, they were all on the mend, and made full recoveries.

The Incas Prime made no difference to the standings, for two Fords beat two Triumphs and *all* of them lost more than 80 minutes. At the end of the South American section, Fords lay first and second, with the Triumphs third and fourth. Only 26 cars were still running. Mikkola's Escort was well clear, but Culcheth was 18 minutes behind Aaltonen's car, and raring to go. As he came down to breakfast, in his Panama hotel, I saw the broad grin, the rub of the hands, and a breezy remark: 'Morning. Nearly finished now. Only another 51 hours to go!'

In fact it was an eventful 51 hours, too, for on the penultimate stage Hopkirk's Triumph was fastest and Aaltonen was pressured so much that he went off the road for 93 minutes, which let Culcheth's big Triumph move up to a magnificent second place, and on the last Prime of all (the 'Aztec' in Mexico) Brian set fastest time:

I certainly think I drove as well as I had ever driven in a rally up to that point. I think it was my best performance—the car was very much to my liking.

WRX 902H, lightened and running with triple Weber carburettors, was Brian Culcheth's victorious car in the Scottish rally of 1970. The mechanics work on the car, Culcheth checks the stiffness of the rear dampers, Johnstone Syer looks cynical, and Peter Browning (far left) is amused. (John Davenport)

Outside the Olympic stadium in Mexico City the two surviving Triumph crews had cause for celebration. Two out of four 'works' cars had finished—a 50 per cent record, in an event where less than a quarter of all starters made it to the end—and finished splendidly, second and fourth. Culcheth had put up best performance on the Panama–Mexico section (which put more prize money into the kitty).

Peter Browning says:

> Nothing went wrong. The service plan worked, the cars worked. Paddy's car was handicapped by overweight by being three-up. If Paddy had been two-up, who knows? He might have outpaced Brian?

In fact, everyone *should* have been delighted by this performance. But they were not.

This 2.5PI started life as a pre-World Cup rally test car, but immediately after that Marathon Brian Culcheth drove it to outright victory in the Scottish rally of 1970.

Lord Stokes's inexplicable reaction was that Abingdon had failed—he simply couldn't gauge the enormity of what had been achieved. George Turnbull was more understanding, especially after Peter Browning sent him a detailed report of what had happened.

> It was our failure to win the World Cup, I think, which signed the death warrant of the Department. Second in the London–Sydney Marathon of 1968 (in an 1800) had been pretty good, *but we hadn't won*. I think that losing again was just too much for him to bear, especially after all the bills came in, and we had 'only finished second'.

Not even a victory for Brian Culcheth on the Scottish rally would do—using the old World Cup rally test car (WRX 902H), which had not only been refurbished and extensively lightened, but was also running on triple Weber carburettors:

Brian recalls:

> It was a lot more powerful than a standard rally car with about 20 extra bhp. At that time Triumph North America were getting a tremendous power output from their PI engines, supposedly over 200 bhp, but the Weber engine we used had about 165 bhp at the flywheel, whereas the World Cup cars gave about 138 bhp at the same point. The Scottish car also had those marvellous gear ratios.
>
> The Scottish car, in fact, was a lot quicker than the World Cup cars, which were a lot heavier—on them it was only the downhill bits which were exciting!

Culcheth's Triumph victory in the Scottish, incidentally, was at the expense of his British Leyland rival, Paddy Hopkirk, who took second place in the ex-World Cup Mini 1275GT.

Searching for new rally cars—then Closedown

After the World Cup rally, Peter Browning was invited to look once again at British Leyland's new-model plans, and to suggest a strategy for the future:

> I don't think we could have made the 2.5PI any better. I think it was a super 'rallying tank', for Marathons and Safaris, but I don't think we could have made it into an outright winner in other events. The whole department, at that time, was becoming like a rather outclassed sportsman coming to the end of his days, and having to decide whether to retire or drop into the second league.

The Scottish-rally winning 2.5PI of 1970 ran in Group 6 ('prototype') form, which explains why it had no bumpers on this occasion. (Brian Culcheth)

In 1970 British Leyland's competitions manager, Peter Browning, was asked to survey the range, and suggest cars which could be developed as motorsport winners. Triumph was not very interested in producing limited-production specials, but a 'hairy' Rover 3500S like this was seriously considered for a time.

Brian Culcheth, as the team's committed long-term professional, had other views:

> I thought there was a tremendous amount that could be done. The 2.5PI had all the potential of the Peugeot 504s, which were similar products in many ways. It could have been so right for all the rallies the 504s went on to win. I was really looking forward to 1971, for Peter was hoping to do the 1971 Safari with World Cup-spec. cars—we were confident about that, for the 2000 already had a reasonable Safari history with privately-prepared cars.

After the World Cup rally, however, where the entire workforce had been out of the Abingdon workshops for many weeks, there was nothing to return to, no new business, and no future plan. Browning now admits that he and his colleagues had been completely bound up in World Cup work, and that nothing was planned, or authorized, for the future.

> I couldn't see anything in the new model 'cupboard' which really would have made a winner. But that wasn't all—it was the failure to get any long-term commitment from *anybody* which was the most frustrating thing—no-one seemed to understand the planning that was necessary to succeed in motorsport.
>
> No-one seemed to understand what rivals like Ford or Lancia were up to, in terms of new models and special cars. They simply couldn't understand—or *wouldn't* understand. I remember many discussions where I would say: 'We should decide on a European rally championship pro-gramme—a series of events,' to which the reply would be: 'We should go individually to events where marketing interests are particularly strong.' They wanted to play around, to *play around*, with different models, and I was sure that was the wrong way to go.
>
> There was no way that I could get through to management that what we had to do was to look at something like the new Triumph Dolomite, and to build in to it, at the design stage, the options which might be necessary to build a future competition car.
>
> Even so, we went to talk to Triumph and Rover people. We got quite a long way down the line, at one time, with Rovers, to actually build a sort of 'Super Rover'. I think we needed one thousand cars, with certain special items installed, like limited slip differentials, flared wheel arches, lightweight panels and things like that. It was my proposal that we could sell a thousand to Leyland dealers around the world, a prestige high-performance version at an inflated price, which would go towards the inconvenience of producing this thousand cars.

Having got those thousand down the line, for homologation purposes, we would then be able to prepare some of them for rallying. Some people at Rover got very excited when Bill Shaw's Rover blew everything off in the first 15 hours of the 1970 Marathon de la Route at the Nürburgring. We wanted aluminium doors, aluminium boots and bonnets—we wanted to lose weight. But that was the closest we ever got, to developing a competition car, a 'homologation special', but the idea was shot down because Peter Wilks was not enthusiastic.

At Triumph I'm sure we suggested playing around with all sorts of things, but Triumph was not amenable to any disruption of the production line. What we should have been striving for at this time was to develop special limited editions from which a competition car be developed. It was really no good trying to compete against Escort RS1600s and Renault-Alpines with tarted-up production cars. Strangely enough, there was more enthusiasm at Rover to do that sort of thing than there was at Triumph.

It was all to no avail. By the end of the summer British Leyland had decided to close down the Abingdon Motorsport operation. On the other hand, Special Tuning, set up in 1964, had expanded mightily in recent years and, with Basil Wales in charge, was making healthy profits. Nothing Peter Browning could say could persuade Lord Stokes to change his mind:

My argument with Lord Stokes was that he wanted to keep on with Special Tuning, and he felt he could continue to be successful with Special Tuning running the cars, and he didn't need a Competitions Department spending all this money. Special Tuning was going ahead by leaps and bounds, it was extremely successful, and was making a lot of money.

My point was that he needed a Competitions Department, for that was his flagship for the sales, developing and testing of Special Tuning parts. I said that, excluding Marathons, that the annual motorsport programme was self-supporting, supported by profits from Special Tuning.

But Lord Stokes said: 'I don't need that. I can make Special Tuning successful, I can have the income from Special Tuning, and I don't need you going off to win the Scottish rally, or what-ever.'

I felt I was going nowhere, and I could see no future commitment, so I just decided to resign.

The news officially broke in August 1970. Peter Browning left the company, and the close down came at the end of October 1970.

Brian Culcheth was in Australia when he heard the news:

I wrote various letters to Lord Stokes, and said that apart from the tragedy of shutting the competitions department I thought it was extremely short-sighted that British Leyland was willing to lose the skills of this department. But I never got very far. . . .

In the next few months, almost all the ex-works Triumphs were sold off, with only one car—Brian Culcheth's successful ex-World Cup 2.5PI—being retained. This, at least, was a blessing, because it gave Culcheth a life line in 1971.

Triumph in the USA—a parallel run-down

Over in California, too, as the competition hotted up and costs increased, British Leyland's commitment to motorsport also sagged, and 'Kas' Kastner became progressively more disillusioned. In 1968 and 1969, however, his major problem had been to make the TR250 and TR6 cars competitive in SCCA racing:

At the time I had a genuine budget, but that might only have been about $100,000 for the whole year, which included two or three salaries for people who were working for me, and the cheques we sent out to SCCA winners.

The SCCA really hurt us by moving the new TRs up a division into Class C where we had to compete with much faster cars, which sometimes outstripped us by 100 bhp. Because they wouldn't allow us to run with fuel injection on the TR250s—but that changed with the TR6s—there was almost no way that we could be truly competitive when we had to run against the

Category C Production mixed an amazing variety of race cars in early-1970s SCCA racing—this being Bob Tullius's Group 44 TR6 ahead of a Porsche 914/6.

Datsun factory team—a *real* factory team that was well supplied with money—and a Porsche factory team which included Richie Ginther.

The Datsuns with the overhead camshaft engines, well, they had valves the size of the butterflies on my SUs! Our TR250s came with two Strombergs, which were anti-smog carbs, but the Datsun was built to win from the start, with three carburettors.

The TR6 had to run in the same category, but I was having to run against Porsches which were so much more sophisticated. $25,000 cars were running against my $6,000 car—there was an astronomical difference in prices at the time.

But eventually I was allowed to use injection, even though it wasn't included on the cars sold in the USA. I'd been lobbying to get the TR6 back into Class D, to replace the TR4, and to run

After a lot of effort by 'Kas' Kastner, the Triumph TR6 became an SCCA-winning car in Category C in North American racing. Bob Tullius was the driver of this car—he later went on to run the Jaguar racing effort in the USA.

Four Spitfires ahead of an MG Midget, with another Spitfire in sixth place—this being a regular scene in North American SCCA racing at the end of the 1960s.

with two Stromberg carbs, but they wouldn't allow that, but they let me run Class C with injection.

Then I found an SCCA rule—it's still in the book, I think—which said you could use any injector, and fuel line, and you could also use trumpets. Great—before long my TR6 engines had trumpets, and the trumpet came with a fuel injector in it! I left the standard injector in the standard manifold, but it wasn't connected to anything.

My injectors were 14 inches from the butterflies, and after making countless tests we had 255 bhp at 7,500 rpm from a 2-litre engine.

'Kas' also had trouble with the Spitfires and GT6s, but made them both work well in the end:

The Spitfire Mk IV engine was a piece of trash, and the 1500 engine wasn't much better. The Spitfire Mk 3 engine was the last good engine they made for that car, it was a terrific engine.

The later GT6 rear suspension, well that made life a lot easier, but it kept failing—we had to make all new parts. We never broke the rubber joints—they took the shock loads off the differential. The parts we always broke were the rear hubs—I finally ended up making my own forgings.

During the last year or so, I had a few links with Triumph in the UK—they wanted me to send over camshafts and things, and I was happy to assist. But the last time I went over to the UK was in 1969, when I met Spen King. I thought he was a wonderful guy. He was probably one of the brightest guys I've ever met in this industry, and he was pleasant too.

But before long I could see that this was going to wind down. There had been a pride at Triumph that I didn't see in British Leyland. Politically I've always been astute, and I could see what was happening.

So I told Mike Dale, who was at Jaguar Inc. in New Jersey, that I was ready to quit, and he said that he wanted me to be competitions manager of the whole outfit. Although I advised Mike about the tactics of running E-Types in SCCA—if the vee-12 was coming, we'd better race the old car so that we could get it into a slower category (Class C) so that when the vee-12 arrived it wouldn't have to race in the top category, but in Class B, which is what happened, and it worked out fine.

In June or July of 1970, though, I gave him 90 days' notice that I was going to quit, and at the same time an old friend of mine, John Brophy, who lived 20 miles from me, wanted to organize a

Brian Fuerstenau was one of the most successful of all Spitfire racing drivers in USA SCCA events. This is a Mk 3 model which was controlled by Bob Tullius's Group 44 team.

racing facility. I could see that the British Leyland job would only last for a couple of years at best, before it was over, so I quit.

There was a successor, Jim Coan, who had been my No. 1 since 1966. But they moved the job to San Francisco, where Jim Coan ran it, then the job became an office, then the job became no job at all.

Preparing cars was the only business I knew, so although it was still John Brophy's hobby I

Bizarre? Don't believe it. This was a serious Trans-Am racing project, a much-modified Triumph Vitesse 2-litre, as prepared by 'Kas' Kastner (in sun glasses), and driven by Carl Swanson in 1971. The location is Elkhart Lake. ('Kas' Kastner)

The Kastner-Brophy team prepared several cars for Triumph-USA in 1971 and 1972. The standard of preparation, as ever, was immaculate.

said: 'Let's do it.' My last day at British Leyland was the end of 1970. I still had a contract, like Tullius had a contract, for 1971 and 1972. I did all the tests for Trans-Am racing.

The problem was that Jim Coan, who'd been my second, could never bring himself to travel the two miles from his office to mine. He was 32 years old, and he thought he was king—king of nothing, though, because I took all my equipment away. In fact I offered him a job to come with me. It was pretty tense between us for six to nine months.

'Kas' Kastner's Gardena workshops—in a suburb of Los Angeles—full of SCCA race cars before an important Californian meeting in 1968. Cars carrying the '44' competition number are the famous Group 44/Tullius machines. ('Kas' Kastner)

By 1973, therefore, the Triumph motorsport effort in the USA had also closed down, but there was a big difference. Unlike Europe, it never came back again.

1971

For months after the closedown at Abingdon, there was no sign of any 'works' Triumphs in rallying, and Special Tuning concentrated its affairs on the modification of Minis for customers. Without Brian Culcheth, and a stubborn desire to carry on in rallying, Triumphs might never again have appeared:

> I was a professional rally driver, and this was a very difficult time for me. I was desperate to keep going, and said so to the Public Relations staff at British Leyland, who included Simon Pearson and Mike Greasley. They must have mentioned this to Bill Davis, who was Triumph's top director.
>
> He called me to a meeting, where I found that he was particularly keen to continue Triumph's association with sport. But there wasn't much money.

It was out of this meeting that the short-lived, but very successful Brian Culcheth–Team Castrol operation was born. Some of the money came from Castrol, some from Triumph, and Brian was also retained by British Leyland International for promotional and demonstration work.

> Special Tuning actually prepared and ran the cars—according to the system Castrol and Triumph were just outside customers.

But what a world of difference between 1970 and 1971. Compared with the huge backing for a World Cup programme, this was a tiny operation. For 1971 Brian Culcheth–Team Castrol tackled just three events, there was only one car (his ex-World Cup machine, XJB 305H) with no spare shells, no dedicated practice cars, and very little service.

The gallant 2.5PI was lightened as much as possible (by removing some long-distance equipment) and tackled just three major events—the Welsh, the Scottish and the Cyprus rallies. In Wales the engine suffered from fuel feed problems, while in the Scottish it was holding fourth position overall before Brian uncharacteristically put it off the road on two separate occasions, before finishing tenth. In Cyprus he came agonisingly close to victory,

In 1971 Brian Culcheth set up his own small Brian Culcheth–Team Castrol effort, and used his ex-World Cup 2.5PI to do a restricted number of events. Here the well-used old car fords a watersplash in the Welsh Rally. (John Davenport)

eventually having to settle for second place.

That, however, was the end of the road for the old 2.5PI, as Bill Davis was looking ahead. For 1972, Triumph was about to launch the Dolomite, and he wanted to sort out this new model as rapidly as possible.

1972

The records show that 'works' Triumphs started only three events in 1972, which rather masks the situated, as a great deal of testing, and some fruitless lobbying, also went on. One of those events was in a 2.5PI, the others being in a Dolomite development car.

Brian Culcheth said:

> One thing I discussed with Davis at this stage was the potential of the 2.5PI in the Safari. Incidentally, I had already made myself into much more of a marketing man, and I was having to 'sell' myself and my abilities to justify the reasons for keeping going.
>
> Bill Davis was keen, but he said he couldn't justify the expense out of his budgets. Then I got in touch with British Leyland International—David Welch, who was head of advertising—and they contacted Benbros who had the Triumph distributorship for Kenya. Between them all, they arranged for the car to go.
>
> It was a brand new car, built up at Abingdon, with a World Cup specification shell, but we had a rally of considerable mechanical problems. It started only 100 miles after the start when an antelope went into the front of the car, and the horns went straight through the radiator. There was water gushing out everywhere. I was with Lofty Drews, a local co-driver, so we got some mud and tried to cake it up, but that didn't work properly. Then we saw a privately-owned Triumph 2000 sitting in a drive, and you can guess the rest! We begged the radiator from the owner, and changed it.

Later in the event the gearbox stripped most of its gears, after the half-way halt the rear screen fell out (that had happened several times in pre-World Cup testing, but body shells had never been strengthened to stop it happening), and finally the whole of the rear of the body shell started to break away from the cabin.

Even so, the car staggered on to win its class and finish 13th overall, but although British Leyland professed themselves to be satisfied, Culcheth was not:

One of Triumph's great disappointments was in the East African Safari of 1972, where this 2.5PI gave a lot of mechanical trouble. Brian Culcheth is still convinced that one of a team of cars could have won, outright, in 1971 or 1972.

Brian Culcheth (left) and Lofty Drews were glad it was all over when they managed to get their 2.5PI to the finish of the Safari of 1972, where they won their class in spite of the car's body shell beginning to break up towards the end of the gruelling event. (Brian Culcheth)

> I think we could have won the Safari in 1971 or 1972 with the 2.5PI—for sure it was better than the Datsun 240Z which was winning in those days. . . .

In the meantime, Triumph's new Dolomite saloon was nearing production, and because there was to be an interesting derivative—the Sprint with a 2-litre engine having only one overhead camshaft but four valves per cylinder—Ray Henderson was allowed to build up a development car in Coventry, for possible competition use.

Brian Culcheth was full of enthusiasm about the Triumph Dolomite Sprint, which he first used in prototype form in the Scottish rally of 1972. (Brian Culcheth)

The second appearance for the prototype Dolomite Sprint, CKV 2K, was on the TAP rally of Portugal, where Brian Culcheth set up a number of fast times, but had to retire with broken rear suspension. (Gordon Birtwistle)

There was never any question of using the 'basic' eight valve engine. Early in 1972 Culcheth was invited to test the new 16-valve car (CKV 2K) at Bagshot where:

> It was so superb, compared with anything we'd had from Triumph before. It was quick and, although the handling left a bit to be desired, basically we were all very excited.
>
> I thought this was the car we should always have been developing. If we'd known about it ear-lier, and if the Peter Browning department had only been put on ice, this is the car we should always have been developing. We should have spent all our time on the car.

This car was used twice in 1972—in Scotland, and in the TAP rally of Portugal. Because the 16-valve car had not publicly been announced, the car had to run in appropriate 'proto-type' classes, and Special Tuning's management made sure that the Press never saw what was under the bonnet.

In Scotland, the car was fast, but not fast enough against the 'works' Escorts which won the event, and suffered from weak shock absorbers, so the result—19th overall—was disap-pointing. Culcheth hoped for better fortune in the TAP rally of Portugal, in September, but on this occasion the car had to retire after one of the angled rear suspension links broke away from the axle.

The car had also lost power during the event, but Abingdon did not investigate before it was returned to the factory in Coventry. Gordon Birtwistle, who then tested the car, remembers that it sounded rattly, felt a bit flat, and that when the top was lifted he found that several rocker arms were broken—the car was only using about 13 of its 16 valves.

After Brian Culcheth's appearance in the TAP rally, the Triumph name took a back seat at Special Tuning, and in British Leyland's motorsport operations. First Peter Browning, then Brian Culcheth, had preached the idea of producing a 'homologation special' with the

Although 'Kas' Kastner left the employ of British Leyland in 1970, his new company, Kastner Brophy Racing, held the contract to run 'works' Triumphs in the USA for some time. These two cars are (left) a GT6 Mk III for Don Devendorf and (right) a TR6 for Carl Swanson.

most powerful possible engine in the lightest possible car, but without response. As Brian recalls:

> Bill Davis of Rover-Triumph said that he couldn't commit himself to a 1973 programme, because the 2.5PI was on the way out and the Dolomite Sprint wouldn't be announced until the middle of the year. He said it was all too early to rely on the Dolomite Sprint, which had tremendous potential.
>
> That meant that I went back to David Welch of Leyland International, who wanted to promote what was really a very lack lustre British Leyland image in Europe. He needed to fire up enthusiasm among his dealers.
>
> Eventually we agreed on an international programme, which would promote Morris Marinas. I was given an office in Leyland House in the Marylebone road, with a secretary, and it all went from there. I had to make everything happen. I made my own programme, I was the team driver, I organized the promotional things; everything.
>
> That was the time in which I drove things like locally entered 2.5PIs in Jamaica and South Africa, with parts supplied from the UK—transmissions, suspensions, diffs. and so on.
>
> In one of my early meetings with Alan Edis I tried to get them to put a Sprint engine into the rear-drive Triumph Toledo, which had a very suitable front suspension for rallying, with a spring/damper unit on top of the top wishbone—really almost a strut, like the Escort had. Nobody wanted to know, though. . . .
>
> In the same way, I tried desperately to persuade someone—anyone—to combine the Dolomite Sprint engine with the Marina, and Leyland International applied pressure to try to get a concept car built. John Lloyd at Triumph said he couldn't allocate men to do that, and Basil Wales said he couldn't do it at Special Tuning 'unless someone pays'. It was very frustrating.

In Europe, however, the Triumph name disappeared from the entry lists. Then came 1974, and it was all-change time again.

Rebirth at Abingdon
1974 and 1975

E ven though the Energy Crisis of 1973/1974 hurt every car-making concern very badly indeed, most of them soon shrugged it off, and looked to their future. In many cases there were reorganizations. One of them, at British Leyland, led to the rebirth of the motorsport department, the reappearance of Triumph in 'works' competition, and the first serious use of Triumph-badged cars on the race track.

For that, Triumph could thank Keith Hopkins, who had been its public relations manager in the 1960s, had become a confidante of Lord Stokes, and was soon to become the marketing chief of Leyland cars. Special Tuning had recently moved to a new site at Abingdon, and was looking for work.

Brian Culcheth was called in to talk to Keith Hopkins, recommended the development of the Dolomite Sprint, and a more aggressive attitude to motorsport. Keith listened, decided to relaunch the department and attracted Bill Price back from his job in a BL dealership.

Although work soon began on the Triumph Dolomite Sprint, Basil Wales's operation was still operating under a great handicap—that of marketing and publicity pressures. Without such pressures no-one, surely, would have persevered with the Morris Marina, and no-one in his right mind would have tried to make a rally car out of the Austin Allegro. But this, don't forget, was British Leyland in the early 1970s, an organization which defied all logic.

1974

The Dolomite Sprint had been launched in the summer of 1973, and had already been homologated before work started on two different projects. Special Tuning began work on a rally programme, using a new car (FRW 812L), while Broadspeed prepared for an assault on the 2-litre class in the British Saloon Car Championship.

Only one driver, Brian Culcheth, was contracted to run the new generation of rally car, and he has admitted that this was a very worrying year, for Dolomites appeared six times and never finished. Brian describes the retirements as 'typical rallying problems'—there was a bent steering rack on the TAP rally (a problem later duplicated in rough road testing), a broken suspension ball joint on the Burmah, and a brake problem on the Ypres 24 Hour rally.

In mid-summer there was a change of management at Abingdon, for Basil Wales accepted an internal promotion, to another department, his position being taken by an energetic young man, Richard Seth-Smith, whose mop of ginger hair soon led to him being nicknamed the 'Flying Carrot'. Those who said 'Who?' when Richard was appointed obviously did not know that he had worked briefly with John Sprinzel's organization in the early 1960s, and that he had more recently been involved in marketing and motorsport matters with Keith Hopkins at British Leyland.

All of a sudden, therefore, British Leyland's motorsport operation began to look credible

People, noise, glamour, excitement—and the 'works' Dolomite Sprint of Culcheth/Syer starting the 24 Hours of Ypres rally in 1974, proudly carrying competition number 1. (Brian Culcheth)

and vigorous once again. This isn't meant to be a criticism of the previous management, but even Basil Wales himself would admit that he was a manager rather than an engineer, an administrator rather than a marketing man—and it showed in the way the atmosphere changed.

The first Triumph Dolomite entry to follow Seth-Smith's appointment was in the Tour of Britain, the second of a short series of events which combined racing with rallying (rather like a Tour de France in miniature), and which was organized for Group 1 cars running on RAC-approved road tyres. Since 'FRW' was a Group 2 machine, this gave Bill Price's team the chance to build a new machine—RDU 983M. Brian Culcheth would drive the car, his co-driver on this occasion being Ray Hutton, who was editor of *Autocar*.

[For those who find rally car registration numbers interesting, RDU 983M was to be used by the Abingdon team for three full seasons—until June 1977—while FRW 812L (the Group 2 identity) had a shorter life of only two years. Both cars, however, started a large number of events, though each was reshelled at least once during its career.]

Competition in the 1974 Tour was very hot—notably from two factory-prepared Ford RS2000s—with five races and 12 special stages along its 1,000-mile route. The new Group 1 Dolomite was a rapid machine, fitted with a Broadspeed-built engine and gearbox, for which more than 160 bhp was claimed, though this was perhaps a mischievous Ralph Broad claim, as the car was certainly no faster than the 140 bhp RS2000s.

It was not a happy event for the new Group 1 car, for it suffered a small engine bay fire on one of the early special stages. Nevertheless, he led two other Dolomites (driven by Tony Dron and John Handley), and set more than one fastest stage time before there was another engine bay fire in the Oulton Park race, which caused the car's retirement. The cause was that petrol had been pumped out of carburettor overflow pipes—a failing which was never allowed to happen again.

The Dolomite/Culcheth/Hutton partnership obviously got on well, however, for they were to compete in two more Tours of Britain—and finish second overall on both occasions. For the rest of 1974, however, Culcheth's Dolomite rallying programme staggered from crisis to crisis; even though a lot of work was done to improve the car's handling and traction (new rear axle location helped a lot—this was perfectly rally-legal on a Group 2 car.) This didn't help on the RAC rally when the car hit a log which had just been dislodged by a previous car and was badly damaged.

On the racing scene, however, the Dolomite made a real impact, with cars prepared by Broadspeed of Southam. Broadspeed, led by its energetic founder, Ralph Broad, had originally made its name in Birmingham, by racing Mini-Cooper S cars which were often faster than the official 'works' teams. Then, in 1966, Broadspeed had been contracted by Ford to prepare a series of white-hot Ford Anglias for saloon car racing, and had subsequently progressed to running Escorts.

In the meantime, Broadspeed had moved to new premises in Southam, south of Coventry, had dabbled with turbocharging of Fords, and race-preparation of BMWs—and started looking around for new business. Ralph Broad, who could talk a kettle into boiling if he thought there was business to be won, saw John Barber (Lord Stokes's managing director) in November 1973, convinced him that there was glory to be gained in saloon car rac-

British Leyland struggled for some years to make the 16-valve Dolomite Sprint engine into a competitive F3 power unit. Here Alan Howell (right) and Mike Connor work on this Holbay-prepared example in Tiff Needell's Unipart-sponsored F3 car.

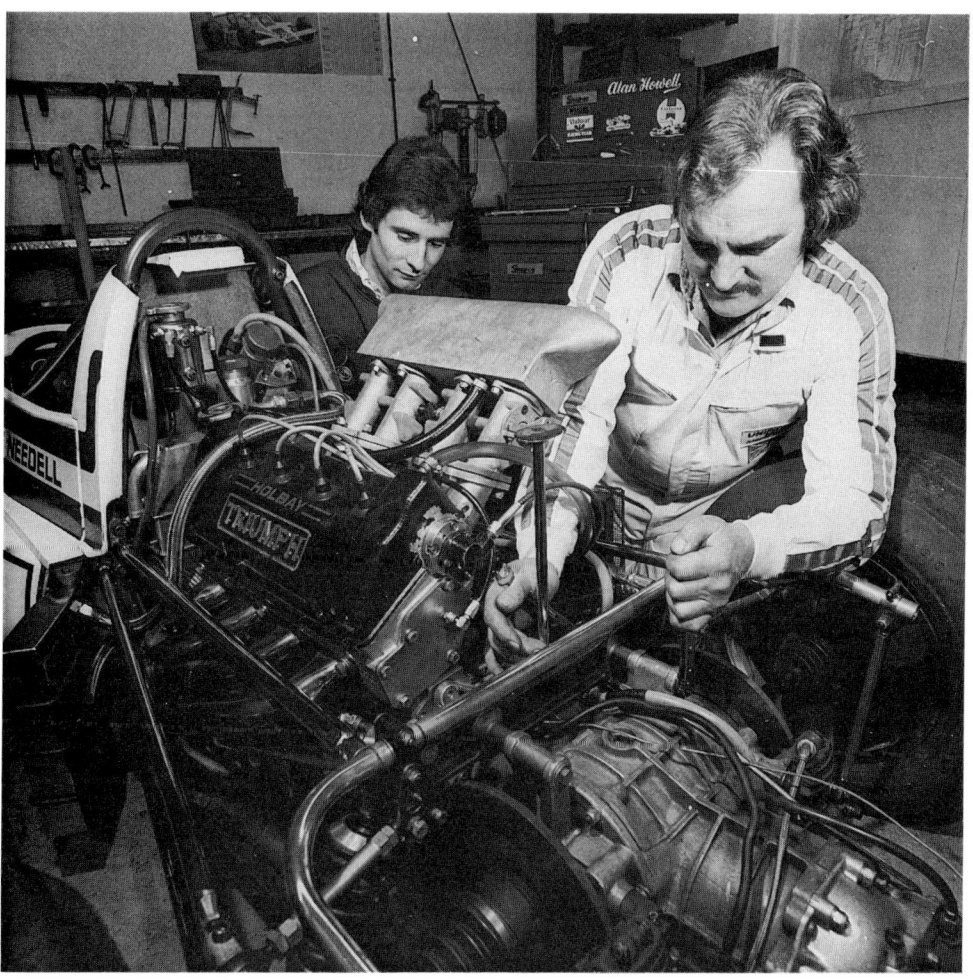

ing, and gained a contract to run a two car team in the British Saloon Car Championship of 1974, the first year in which Group 1 regulations were imposed.

There was very little time to get two cars ready for the British racing season, which opened in March. The contract was only signed in January 1974, and the two cars barely made it to the first race. It is not surprising that many improvements were made in the years which followed. Much of the ongoing development concerned tyres (from Dunlop) and work on suspension settings.

The team's number one driver was Andy Rouse, who worked for Broadspeed at the time, and after a test session at Silverstone journalist Tony Dron (surely the tallest racing driver in the business?) gained the second seat. Broad's engineers worked their magic on the engines, boosting power from 127 bhp (standard car) to 174 bhp; the rev limit was no less than 8,200 rpm—not a bad improvement for Group 1, and the gentle art of blueprinting! Ralph Broad's unique solution to the use of overdrive was to wire it only for use on third gear—except at Spa where it was wired in for use in top gear too.

The car's major problem, at first, was its brakes (the car only had 8.75 in. solid front discs, hidden inside 13 in. alloy wheels), and there was overdrive trouble in the early races, but the cars soon became reliable, even for long-distance events.

Throughout the season, Andy Rouse was always the fastest driver, but Dron got closer and closer to him as he built up more experience with the car. In the end, not only did Andy Rouse win the 2-litre class of the British Saloon Car Championship, and British Leyland won the Manufacturers' title, but Tony Dron finished third overall in the Tourist Trophy race, over 105 laps of Silverstone's Grand Prix circuit, behind two vee-8 engined Chevrolet Camaros. Would you like to tackle a single-handed three-hour stint of Silverstone in a high-revving Dolomite? Tony Dron didn't seem to mind.

In July, when face with the might of 1973-style Group 2 'super-saloons', Rouse and Dron also entered the Spa 24-hour saloon car race. They used the car which Dron had driven in the Tour of Britain, now fitted with dual-choke Weber carbs and overseen by Broadspeed, and pedalled it flat-out through the 24 hours to finish fifth overall, having averaged 102.1 mph.

Clearly there was a lot more performance to be wrung out of the Dolomite Sprint in the future, especially on the race tracks. It was too heavy, a little high-and-narrow, and those brakes were always a worry, but as far as British Leyland was concerned it was a lot more

Brian Culcheth and Johnstone Syer pressing-on hard in the dust of a Scottish rally special stage. SOE 8M had been used as a works-supported car in 1974 (by John Bloxham), but had a new shell for this event. It didn't help—Brian crashed the car into the trees. (Brian Culcheth)

encouraging than dreary machinery like the Marinas and Allegros could ever be! Certainly, with 200 bhp already available from Group 2 engines built by the Don Moore business, it was a very fast car indeed.

By the end of the season, Brian Culcheth was fired with enthusiasm for the Dolomite Sprint, as he made clear in an *Autosport* interview published in the autumn of 1974:

> I really do think we are getting to the stage now when the car is going to become a rally winner . . . I think that our Burmah stage and our Tour of Britain stage and also the times we've done at Bagshot recently indicate that the car is, at last, in the competitive frame . . . I think the Dolomite will form the spearhead of our forest rallying for a while, but we have some really exciting cars coming along that I think will see British Leyland really back on the competitive map in the next year or so. . . .

The last comment, for those in the know, clearly referred to the new Triumph TR7 sports car, which was due to be launched in January 1975.

1975

Even before the motorsport season began, British Leyland hit the headlines in a very big way when it struck a severe financial crisis and had to approach the government for support. The result of this was that within weeks the company was rescued by nationalization, and a wholesale change of personnel eventually took place.

Although John Barber and Keith Hopkins left the re-formed company during 1975, and Lord Stokes became no more than an honorary president, there was no immediate effect on the personalities at Abingdon. Fortunately the projects already agreed were confirmed, with the rally team splitting its time between Dolomite Sprints and Morris Marinas.

In January 1975, too, British Leyland announced the new TR7 sports car, and well before the end of the year it had decided to use this as its new rally car for the late 1970s. The development of this car, and the people who arrived to develop and run it, meant yet another change of emphasis, and yet another upheaval for Triumph's motorsporting history.

As in 1974, the Dolomites had a straightforward job to do—Brian Culcheth was the sole team driver to go rallying, while Broadspeed had a two-car team with which to race in the

On the Lombard-RAC rally of 1975, Brian Culcheth used the hard-working RDU 983M Dolomite Sprint in Group 1 guise, not only winning his class, but also winning the entire Group 1 category. (Brian Culcheth)

British Saloon Car Championship. In each case the Dolomites had the benefit of 2 in. SU carburettors and a new camshaft profile, newly homologated and able to liberate more power.

In 1975 Brian Culcheth started 12 events, one of them being the Tour of Britain, when he was partnered by Ray Hutton and used the same Group 1 Dolomite as in 1974. In the other 11 events he was partnered by Johnstone Syer and concentrated on the British Rally Championship, starting nine times in the hard-working FRW 812L Group 2 car. This pair also appeared once in a new car (SOE 8M), and once (on the Lombard-RAC Rally) in the Group 1 example.

Compared with 1975, this was a much more successful programme. Not only did Brian and Ray finish second overall in the Tour of Britain, but Brian and Johnstone won the Hackle rally outright, and finished third overall twice (Jim Clark and Lindisfarne rallies). Their best result was also their last of the season, when they won the prestigious Group 1 category in the Lombard-RAC rally.

Seventh overall on the Mintex was a good way to break the Dolomite's habit of retiring from events, but an outright win on the Hackle (not a British championship event, admittedly) was even more encouraging. More important, the car was reliable throughout, and Bill Price's team of mechanics beamed. Dogged perseverance was beginning to pay off.

Fourth overall in the Granite City rally, behind three Escorts (the winner was Roger Clark in the brand-new Ford Escort RS1800) boosted morale even further, but an accident in Wales, and another accident (in SOE 8M) in the Scottish brought everyone back to earth in more ways than one. Fortunately Brian bounced right back, with third overall in the Jim Clark rally a few weeks later.

Then came the Avon Tour of Britain, that unique race/rally/endurance event which Culcheth had so badly wanted to win in 1974. This time around, there were five circuits and no fewer than 16 special stages in the event—the balance being thought to favour rally rather than racing drivers. For Abingdon, it was a question of 'same again', for Brian shared the up-dated 1974 car with the same co-driver—Ray Hutton. The event started and finished from Birmingham, the main difference in the entry being that Ford did not enter 'works' RS2000s, but gave a lot of help to Tony Pond in a dealer-sponsored RS2000 instead.

In the end it was Pond's Ford which beat Culcheth's Dolomite by a mere 47 seconds and

Brian Culcheth and Autocar Editor Ray Hutton teamed up to tackle the Tour of Britain in Dolomite Sprints in 1974, 1975 and 1976. This mixed race-rally event included fast special stages like this one, the Eppynt Ranges in Wales, where the Group 1 Dolomite could really stretch its legs. (Brian Culcheth)

Whoever said Group 1 saloon cars were boring? This study of Brian Culcheth's Dolomite Sprint, at Cadwell Park in the 1975 Tour of Britain, nails that lie, once and for all. Brian was on his way to second overall. (Brian Culcheth)

Brian (who went on to manage Tony Pond's affairs in the mid-1980s) commented years later: 'I didn't know much about Pondy until the Tour in 1975, when the little b****r disobeyed orders and beat me!'

The 'works' Dolomite Sprint was an ideal Group 1 car for events where circuit tests and tarmac stages predominated, though Brian always said he thought the car was too heavy, and the brakes were not good enough. With Ray Hutton as his co-driver, however, he finished second overall in this, the 1975 Tour of Britain.

It was an event which should have favoured the racers, because there were two hours of circuit racing to only 45 minutes of special stage motoring. Clearly Ray Hutton enjoyed himself enormously, heading his *Autocar* report: 'Rallymen race to the finish' and pointing out the good-natured badinage which took place between Culcheth and Pond:

> It was 4 o'clock on Saturday morning at an inhospitable disused airfield called Stoke Park. Tony Pond's two-tone blue Escort RS2000 was parked alongside Brian Culcheth's red, white and blue Dolomite Sprint, waiting to go on to the special stage.
>
> 'I suppose there is nothing seriously wrong with your car?', Culcheth asked, hopefully.
>
> There wasn't. . . .

Hutton also let slip the way in which the latest items had been homologated:

> The Dolomite's new camshaft, which totally destroys the smoothness and refinement of the simple 'blueprinted' engine we had last year, in the interests of a big burst of power at 5,000 rpm and beyond, comes as part of—believe it or not—an 'emissions control kit'. It will not have escaped readers' notice that the Dolomite Sprint is not sold anywhere that very strict exhaust pollution laws apply.

Brian was so wound up at the after-rally prize presentation (shall we say that he had lubricated himself rather well?) that after he had been given a celebratory birthday cake he squashed it all over Ray Hutton's face.

In the Tour, Culcheth's car was very fast on the circuits, winning outright on occasion, but it could not match the RS2000 on the stages. Even so, it wasn't all plain sailing for the Dolomite:

Brian says:

> On the Tour of Britain we were always quick on the circuits, but the engine always overheated and we lost power. Tony lost an engine that way in the 1976 Tour. I didn't have one fail, never, mainly because I nursed mine a bit, and I always raced with the heater full on. The brakes on the car were always poor. British Leyland would never homologate anything. There was still a long way to go with that car when we stopped using it.
>
> We never developed the Sprint to its limit. I drove RDU again in 1991, and I've decided that one of the things which would have made the car much nicer would have been power steering. It all came back to me, the aches in the shoulders, everything—it was never a car you could sit back and play with, not the way you could play with an Escort. You were on a knife-edge all the time, and had to grit your teeth to make it work.
>
> We struggled with the gearbox for years. It took me five years even to discover that the Scottish rally 2.5PI had the special ratios of the mid-1960s 2000s. Those ratios still existed, they would have fitted the Dolomite Sprint gearbox; that was one area which could have been improved.

After the excellent performance on the Tour of Britain, the old Group 2 Sprint took third overall in the Lindisfarne rally, but then it was Lombard-RAC rally time, and BL decided to attack the Group 1 category. Once again the Tour of Britain car was refurbished, in yet another type of livery, and Brian and Johnstone duly delivered the goods.

In an event dominated by the battle for the lead between the Group 4 'works' Escorts and Lancia Stratos models, the Group 1 category never made many headlines, but the Sprint completed the difficult route in 16th place overall. More important, it not only won its 2-litre capacity class, but it also won the Group 1 category outright. A year on from the 1974 event, when the same car/crew combination had been put out in a bizarre accident, this was a great advance.

Once again Broadspeed was contracted to run two 'works' Dolomite Sprints in the British Saloon Car Championship. For 1975 it was a two car programme, for Andy Rouse and Roger Bell (Editor of *Motor*) to drive, with the cars handling better owing to a revision in

The 'works' Dolomite Sprints were used in both Group 1 and Group 2 guise. The way to 'pick' a Group 1 car—like this—was to observe the standard wheels, and the lack of wheelarch extensions. Group 2 cars usually had flared wheel arches to allow wider wheels and tyres to be used.

suspension geometry, and because longer turret-type rear dampers were fitted. The engine had been made considerably more powerful than before—peak power was up to 185-190 bhp (in a Group 1 car? No wonder people accused Ralph Broad of wizardry.)—the transmission was made stronger (particularly the back axle), and the car seemed to handle a lot better than in its first year on the tracks.

Once again Rouse proved to be the faster of the two drivers, and at the end of a long, hard-fought and tense season he not only won the 2-litre class, but he also won the Championship outright. This is not to say that the Dolomite was the fastest car on the track—that honour usually went to one of Chevrolet Camaros driven by heros like Stuart Graham, Richard Lloyd or Vince Woodman—but Rouse was the most consistent of the class winners, which gave him the title.

This season there was no money available for Broadspeed to build a special Group 2 car for the Spa 24 Hour race (though Ralph Broad was convinced that he could win the race, knowing what he had learned in 1974). Instead, all the same ideas were fed into the building of a 235 bhp-engined example, which was built to European *Trophée de l'Avenir* regulations—a pseudo Group 2 formula adopted for the International Tourist Trophy race at Silverstone, and one which was forecast for adoption as an official Group 2 formula in 1976.

This was a superb-looking machine, newly-liveried in blue-and-white British Leyland colours—the first time that the name of the government-owned 'Leyland Cars' organization had appeared as overall sponsors of a competition car. The 235 bhp power output (at 7,750 rpm, from a 2-litre engine, remember) was achieved by using two vast dual-choke Weber carbs, free-flow exhaust system, a modified rear suspension (with Panhard rod linkage like that of the Abingdon Group 2 cars), much larger front disc brakes, of 10.3 in. diameter, and

The Broadspeed Dolomite Sprint race cars were always beautifully turned out—and were winners too! This is the car driven by Andy Rouse, which won the British Touring Car Championship outright in 1975. Posed alongside, at this exhibition, is Sir Ronald Edwards, who was chairman of British Leyland at this time.

9 in. rim alloy wheels. Andy Rouse was given this car to drive, while Roger Bell/Jenny Birrell had a more normally-specified British-Championship Sprint.

Broadspeed had to argue its point about the eligibility of the Panhard rod with the event scrutineers, but Rouse's car was so fast that it shared a grid front-row slot with pole-sitter Stuart Graham (5.7-litre Chevrolet Camaro); Roger Bell's car was five seconds a lap slower on Silverstone's GP circuit.

In the early stages of the race, Graham's Chevrolet led, with Richard Lloyd's Camaro close behind, but Rouse was comfortably in third place, and the crowd loved the Dolomite's giant-killing act, while it lasted.

But it didn't last. The Dolomite contacted the Camaro on a corner, and needed a pit stop to have the wing pulled off a wheel, dropping to 24th place. After an hour it was catching up well, and the Bell/Birrell machine was fourth, but both cars then struck trouble. Rouse's car suffered a worn-out gearbox, while the other Dolomite suffered a persistent misfire, and dropped back to fourth place in its class.

In the race, however, Rouse lapped in 1 min. 45.64 sec., which equated to 99.92 mph, and was easily a 2-litre lap record for the circuit. If the 1976 cars were to be close to this performance, they would set new standards in saloon car racing.

Big changes ahead

Well before the end of 1975, British Leyland announced a major expansion in its motor-sport programme. Press releases published in October 1975 gave the impression that the company was re-opening Abingdon, and a 'works' team, for the first time since the World Cup of 1970—but if so, what had all the Dolomite activity of 1974 and 1975 been? A mirage?

As far as the enthusiasts were concerned, however, this didn't matter. What *was* important was that the company, now state-controlled, was making a major commitment to its sporting future. Both in motor racing, and in rallying, the chosen cars were Triumph-badged. The use of Dolomite Sprints on the circuits was an obvious solution—but the choice of TR7s to go rallying was a real surprise. In 1976, and in future years, exciting times lay ahead.

Chapter 9

Wedge styles and vee-8 engines

World Championship outings, 1976-1980

Although cynics might not agree with this, when British Leyland began to put more backing behind Abingdon in 1974 the marketing specialists *had* a master plan. Class wins, and 'good show' appearances were not enough—for 1976 and beyond, the company was shooting for outright wins.

Even before the end of 1974 Richard Seth-Smith arranged for Bill Price and his team to have an early look at the still-secret TR7 sports car, which was due for launch in January 1975, to assess its potential as a rally car. Price, initially at least was not impressed:

> The car seemed far from ideal in many areas, with similar rear axle location to the Sprint, a short wheelbase, a single OHC engine and it was rather heavy. The wedge shape, with the possibility of homologating the right parts, plus the knowledge that a vee-8 engine was coming, made this model about the only one which had a chance of being developed for rallying.

Early in 1975 Abingdon took delivery of an old TR7 development car from Canley, and started development work. It wasn't long before Price, Brian Culcheth, and the technicians decided that a great deal of work would be needed to turn it into a winner. The original intention was to run it as a 1,000-off Group 3 car (the 500-off Group 4 category came later). First of all it would have to be made faster than the existing Dolomite Sprint, and only then could sights be set on matching the performance of the 240 bhp 2-litre Ford Escort RS1800s.

It was here that Bill Price's famous 'inventive homologation' skills came into play. By a careful reading of FIA regulations (and, perhaps, taking lessons from Ralph Broad, who was an old hand at the black art of 'reading the rules') Bill managed to gain approval for a TR7 which was only loosely related to the production car.

Instead of the standard car's ohc 8-valve engine, he got the Dolomite Sprint's 16-valve engine approved. Instead of the standard car's four-speed gearbox, he got the Sprint's more robust gearbox *and* its overdrive approved. He also got the heavy duty rear axle approved as an option before it had truly gone on sale.

The fact that no TR7 production car was ever sold to the public with a 16-valve cylinder head or with an overdrive gearbox didn't stop the car from being homologated. John Davenport remembers that the '100-off' rule applied to alternative engines at the time (which explains why Ford was able to race 24-valve four-cam Capris in 1974), but this waiver never applied to transmissions. In any case, the long-term ambition was also to use the new five-speed gearbox which was being developed for the Rover 3500 and the TR7, but this was not available to the team until the autumn of 1976.

The TR7 was actually homologated on 1 October 1975, and British Leyland's intentions became clear later in the month when it announced its 1976 programme. A continuing Broadspeed racing effort with the Dolomite Sprints was confirmed, and, to paraphrase one media report:

The rallying programme centres around two 16-valve, 5-speed Triumph TR7s, to be driven by regular Leyland driver Brian Culcheth and former dealer Opel Team driver Tony Pond. The cars will be developed and prepared at Leyland Special Tuning at Abingdon, and will be overseen by Bill Price, ST's workshop manager who was with Abingdon in the glorious Mini days.

Overall, Leyland's plans are both enterprising and exciting. The company has hired an excellent team of drivers, can depend on first-class preparation, and its involvement is sufficiently diverse to get the Leyland name across to a wide variety of spectators and media.

Culcheth, of course, was an ideal team leader, while Pond was a real 'capture', for he had become a very hot property in the previous year or so. Culcheth, in particular, was not likely to forget being beaten, so narrowly, by Pond in the 1975 Tour of Britain. Brian says:

Things began to hot up in 1975. Leyland House said 'we can't be seen to muck about anymore', and they needed another driver to join me, preferably British. We talked to Russell Brookes, Will Sparrow and Tony Pond—we tested them all at Bagshot in a Dolomite Sprint. We came to the unanimous conclusion that Pond was the man for us. He seemed pleasant enough to get on with, and we felt his driving was the best of the three.

But these were the professionals being guided by well-meaning amateurs. Unfortunately for the professionals, who had to deliver the goods, this was the start of an absurdly vainglorious period in Leyland's publicity and public relations areas. There was a great deal of glamour, of glitz, and an impression that outstanding success was not only just around the corner, but guaranteed. When this proved not to be so in 1976, the media reaction rapidly turned sour. Peter Newton, who was *Autosport's* Rallies Editor, summed up at the end of the 1976 season in this way:

Leyland began the season with disaster and ended it in triumph. Having announced their TR7 programme prematurely at an ill-judged Press reception in the autumn of 1975, the eagerly-awaited cars finally appeared for the Welsh in May and were almost immediately relegated to the ranks of also-runs . . . As the months had gone by, it had become apparent that Leyland's new TR7 programme was nowhere near as staunchly supported by the various institutions within the Corporation as had been intimated by the size of the investment. It became clear that parts supplies were at best sluggish, and that 'one-off' orders were being treated with no apparent urgency whatsoever.

Motoring writers were not amused by the fact that cars homologated in 1975 were not ready to be rallied until seven months later, and did not even finish an event until ten months had passed. If there had been no victories before the end of the first season Abingdon might have been a laughing stock.

1976

Although BL's motorsport year started with Dolomite Sprints, TR7s, and Austin Allegros all being prepared for entry in rallies, the unsuccessful Allegro was dumped in March. As soon as possible Culcheth and Pond concentrated on the TR7s, and a Group 1 Dolomite Sprint was entrusted to a new recruit to the team, Pat Ryan.

Early in the year the Abingdon team realized that they would be hard put to get two TR7s to the start line of a rally, let alone have them competitive by that time. Plans to start the British Championship season with TR7s on the first event were soon thwarted, and it was not until May—the Welsh rally—that the cars actually appeared. In the meantime Culcheth and Pond 'made do' with the existing Dolomites—Brian driving the Group 2 car, and Pond the Group 1 car.

The problem was an ongoing one—that Triumph would rather not have been part of British Leyland at all, and would have preferred to run its own motorsport programme. The engineers were always reluctant (or instructed to be reluctant) to help the Motorsport department at Abingdon.

The first TR7 rally cars used 16-valve Dolomite Sprint-type engines, and ran with twin dual-choke Weber carburettors. The brace between the strut towers was used to add overall rigidity to the body shell.

Bill Price's brief was to involve Triumph, in Coventry, as much as possible in development, and always to buy his special parts through the engineering department's existing system. In Gordon Birtwistle's succinct words, there was again a great deal of 'them and us' in the relationship, and this attitude applied in both directions. Brian Culcheth recalled: '. . . a tremendous political infight between Coventry and Abingdon, which was going on all the time; it was unbelievably frustrating.'

The first TR7 test was delayed until March 1976, but Leyland public relations staff then made the big mistake of inviting the British motoring Press along to Bagshot to ride in the car when it was only on its third test session. Impressed by the sheer exuberance and bravery of the team's drivers, the Press greeted the experience with enthusiasm—and suggested that it could start winning very soon.

Even after so much development, however, close-ratio gears for the new five-speed gearbox were not available, the car still had to run on rear drum brakes, and the standard rear suspension linkage was retained. The twin-Weber'd engines, at least, were high-revving and powerful, so the car was very fast in a straight line.

The Dolomites started their season in the Tour of Dean, where Tony Pond won the Group 1 category in RDU, in his first drive for the team. On the Scottish Snowman rally (which *was* very snowy), the Group 1 car blew its engine, while Culcheth's Group 2 car finished fifth overall. Then, on the Mintex, held in Yorkshire, both cars finished in the Top Ten, with Tony Pond once again winning Group 1. In Scotland, on the Granite City, Pond then took fourth place in the Group 1 car and won the category, which was a fabulous result.

Things could have been worse; but not much. When two TR7s turned up at the start of the Welsh rally in Cardiff in May 1976, they were still lamentably under-developed, and they had had to be fitted with Dolomite Sprint-type gearboxes and overdrives. The miracle was that the cars had come so far in the time available, and were even competitive.

Brian Culcheth was particularly disappointed:

> Tony and I had done the majority of the testing, and although there was no fundamental problem, there were a series of silly mistakes. We knew the engine was always vulnerable to head gasket problems, and we had trouble even in getting the gaskets made of the right type that we knew would work.
>
> We had tremendous chassis problems with the car at first, because we simply couldn't make it work properly. The TR7 frightened me to death sometimes, it was an extremely unpredictable car on loose surfaces. You could go testing with it, and you could go round the same corner three times and get the same effect; the fourth time you would do exactly the same thing, and it would go off into the trees.
>
> We couldn't really figure out why this was so. Certainly the suspension movement was very limited, but I think it was just one of those concepts which was not good for forestry rallying. It was great on tarmac, but in forestry stages—it was either too wide, or too short, or whatever.

In the 1976 Snowman rally, British Leyland was still waiting for its TR7s to be finished off, so Tony Pond and (seen here) Brian Culcheth used Dolomite Sprints instead. Using FRW 812L, running without a front bumper, Culcheth took fifth overall. (Hugh Bishop)

In its first year the TR7 struggled to be reliable and competitive. Tony Pond, here seen on the Lindisfarne rally where he finished ninth, was one of the two regular drivers.

But I *do* think it got better when it became a vee-8. From what Tony told me later, it handled better, and was a lot more predictable.

The biggest controversy, however, centred on the engine:

The car wasn't particularly quick in a straight line. I don't think the engines seemed to produce the power that we'd already had in the Dolomites. We lost a lot of flexibility—there were various problems with the cylinder block, which kept breaking under high revs. In the earlier [Dolomite] programme we never got over 7,000 rpm—the Don Moore engines were superb.

Then, in the TR7s, I had Don Moore engines and Tony had Broadspeed engines. Mine were considerably more torquey than Tony's, though he probably had a bit more top-end power. Andy Rouse and Co. had been using more than 8,000 rpm in the Dolomite race car programme, and we started being told to use more revs too. There was a great deal of pressure on us to use the Broadspeed spec.

But we had many other problems—suspensions, axles, transmissions—I remember using three gearboxes on the RAC rally in 1976.

In the Welsh rally both cars broke their engines during the first night; both blew their cylinder head gaskets, Culcheth's on the first special stage of all, Pond's a few hours later. A few weeks later, on the Scottish, two more engines failed, and even an illegal (in sporting terms) engine change came to nothing when the replacement exploded on a high-speed special stage. I was observing on special stages at that event, and winced when I heard the spectators jeering at the misfiring cars as they crawled past.

Poor Pond even suffered an engine failure in a Dolomite Sprint in the fourth and last Tour of Britain, though Brian Culcheth finished second yet again. For Brian it was almost a repeat run of 1975, except that the Escort RS2000 which beat him was a Mk 2, driven by Ari Vatanen.

Brian's car MYX 175P—a Group 1 example, don't forget—had 178 bhp, and was fastest on many of the stages (there were 26 of them on this occasion), but could not always keep

The interior of a 'works' TR7 looked like this in 1976. At this stage the overdrive transmission was still in use, which explains the switch fitted to the gear-lever knob.

up on the circuits. It didn't help his pride when Vatanen inadvertently punted him off the road in the Silverstone race but he never gave up trying. This time the losing margin was 52 seconds.

It was time, surely, for the TR7s' luck to change, and this happy event duly took place on the Manx international rally in September. Significantly, this was the first tarmac event that the TR7s had tackled, and it was immediately clear that they were always likely to be more effective in these conditions. Let's forget all the nonsense talked about TR7 problems being related to 'poor driver visibility'—if that had been so, on the high-speed stages of the Isle of Man, how could they ever have achieved such good results?

The first few words in my *Autocar* report on this event were:

'Ford versus Porsche' should have made all the Manx headlines last weekend, but it didn't. The real news was that Leyland's works TR7s came good—very good.

Almost everybody, and especially Abingdon's staff, was relieved. The TR7s first failed, then finished slowly in previous events, but all the signs are that their nightmare is over. Tony Pond's car took third place and Brian Culcheth's car fifth. Leyland's weekend was completed by Pat Ryan winning the Group 1 category in his Dolomite Sprint.

The three cars also won the team prize, and in spite of the fact that Pond had spun at high speed, rattling the car against the banks, and damaging the front end, everyone was delighted.

To regain all its credibility the team needed a victory. Tony Pond repeated his Manx third place on the Castrol 76 rally in mid-Wales (this event was already a traditional 'pre-RAC rally' test event, where other works cars sometimes turned up), and both the TR7s felt, and went better because they had abandoned all Broadspeed bits in favour of Don Moore bits, and were more flexible as a result.

Then Pond was sent to tackle a less-important national-status event, the Raylor rally, which was held in Yorkshire, and finally achieved the victory the team so richly deserved. His usual 1976 car (KDU 498N) was not only fitted with the latest parallel-link rear sus-

KDU 497N, one of the original 'works' TR7 rally cars, had a long and hard life, usually driven by Brian Culcheth and Johnstone Syer. The scene is a ford on the Castrol 76 rally, where Brian took eighth place. Teammate Tony Pond finished third in KDU 498N.

pension, but the Abingdon-developed high-geared steering rack, and revised geometry. It looked a better car—it *was* a better car.

The last event of the team's season was the Lombard-RAC rally, which started and finished in Bath. Although two brand-new TR7s were prepared for the usual drivers (though

The 'works' TR7's first victory was in the National-status Raylor rally of October 1976, when Tony Pond used KDU 498N in a test and development exercise before the Lombard-RAC rally.

Pat Ryan had to soldier on with the incredibly hard-working Group 1 Dolomite Sprint, RDU 983M), only one of the cars completed the event.

Pond's TR7 started very rapidly indeed, mixing it with all the fastest Escort RS1800s, the Lancia Stratos, and the Saab 99s, but was eventually forced out when the rear suspension radius arm locations broke up. Culcheth's car suffered similar problems (which were discovered, and rectified), but his biggest problem was with the new five-speed gearboxes:

> We got through three gearboxes. We got back to Bath on the first loop with only fourth gear left, but somehow we managed to crawl back. The next day we put a new box in—it lasted two stages, then it lost first and second gears. Then we had to fit another box, and only three speeds were still working on our return.

Ninth place overall, 20 minutes off the pace, but with several 'works' Fords, Saabs and the Stratos in front of the TR7, showed how hard Brian had worked to stay in the event.

However, as Peter Newton's end-of-season *Autosport* survey pointed out, the team's problems had usually not been of their own making:

> To enter RAC rounds for development purposes having announced the programme with such confidence and forcefulness was to invite disaster. Appearance deadlines came and went, and a 'full assault' on the championship became some 'selected rounds for development purposes'.
>
> Perhaps Leyland would have been better advised to have entered less sensitive areas of competition at the outset . . . The gradual revitalization of Abingdon throughout the year has been a fascinating if somewhat painful evolution. There is still a long way to go, but the drivers have shown that the car is now competitive . . .

Well before the end of 1976, however, British Leyland had decided to make another change to the way its motorsport operation was run. Until then Bill Price had managed the Abingdon department, reporting to Richard Seth-Smith, but when Seth-Smith decided to leave the company to take up another job, Leyland Cars decided to create a new job—that of Motorsport Director. It was the first time the company had taken motorsport so seriously.

To find the right person Leyland used an outside agency, the result being that John Davenport (press officer of the RAC Motor Sports Association) was recommended for the

In the TR7's first Lombard-RAC rally, of November 1976, Brian Culcheth took this new car (OOE 937R, on its very first event) to ninth place overall.

John Davenport, who became British Leyland's Director of Motorsport at the end of 1976, and who presided over the final seasons of the 'works' Triumphs.

new job. John's comments, made in 1991, confirm that the mid-1970s motorsport effort was certainly lacking in direction:

> I was approached by a headhunter. Officially he came to the RAC Motor Sports Association looking for information on racing and rallying, but the next thing I knew was that I was invited to talk to Alex Park and Derek Whittaker of Leyland, and was offered the job as Motorsport Director.
>
> I was hired before the end of the year, in fact I joined in December 1976, but the 1977 programmes had all been settled by then. One of the first difficulties, which had to be sorted out, was the financing. In the early 1970s a lot of the money had come from Leyland International, then Special Tuning had soldiered on, being passed from hand to hand in the company.

[By 1976 it was controlled by Leyland Cars Public Relations, though much of the money still came from BLI.]

> Then came the Jaguar racing effort, and there was also the Dolomite racing programme, which Marketing was paying for. It needed co-ordination, which it had never had in the past.
>
> When I arrived, I reported to Alen Edis, who was director of product planning, and he in turn reported to Spen King, who was the overall engineering director of Leyland Cars. Spen in his turn reported to Derek Whittaker . . .

John already had a great deal of experience in international rallying (though not, by his own admission, very much in motor racing). Like many of us, John had divided his early 'twenties between rally competition and journalism, had worked for *Motoring News* and later for *Autosport*, co-driven for many of the world's top drivers including Hannu Mikkola and Ove Andersson, and had joined the RAC MSA in 1975.

The first problem, at Abingdon, was to find an office for the new director, this soon being created from a small showroom at the corner of the current motorsport building. For John, this new job was going to be no sinecure. In addition to the TR7 and Dolomite Sprint rally programmes, he also had to oversee the Broadspeed Jaguar and Dolomite Sprint racing efforts, Mini racing, and even the development of the Unipart F3 team, where Dolomite Sprint engines were to be used.

Tony Pond   Fred Gallagher

For 1976 the British Saloon Car Championship was sponsored by Keith Prowse. Engine sizes were limited to 3-litres, which immediately meant that Ford's 3-litre Capri set the markers. Once again, however, Andy Rouse drove the Broadspeed Dolomite Sprint in the 2-litre class, and had to battle throughout the year with Gerry Marshall's thoughtfully-homologated Vauxhall Magnum Coupé. On this occasion the Dolomite was beaten into second place in the class—one result of which was that Ralph Broad, and British Leyland, made even greater efforts for 1977!

The very first F3-tuned Dolomite Sprint engine had appeared in November 1975, but it was not until 1976 that Unipart backed a serious programme. Tony Dron drove the Unipart March, while Holbay tuned the engines, but he had to struggle against unreliability at first, and against the fleet of successful Toyota-engined cars. By the end of the season Unipart had also backed the Anson team, and both had achieved top-six results by the close. To win, though, with this unique engine, would need a redoubled effort in 1977.

1977

John Davenport's first season in control of the 'works' programme surprised him, and brought many detail changes to the way things were done at Abingdon. He was, at least, relieved to be told that it was company policy that the cars should run without commercial sponsorship of any sort—which meant he didn't have to spend hundreds of hours drumming it up. On the other hand:

I didn't think I should be employed as a finance or a marketing director. It was only after three or four years that I began to realize that the amount of time I actually spent working on pure motorsport matters was about 10 to 20 per cent. . . .

The 1977 season was one of almost complete contrasts. On the one hand the TR7 recorded its first outright victory—and was always competitive—while the first vee-8 engined TR7 V8 development cars took to the road for testing. On the other, morale within the team plummeted after the drivers were 'gagged', and not allowed to talk to the motoring press without a 'minder' being present. The result was that Brian Culcheth, who had done so much to keep Triumph in motorsport in the early 1970s, left the team.

Strong personalities, and opinions, were in evidence on all sides, and for me, as a working pressman, it was a prickly period to be close to the team. Brian Culcheth is still bitter about the way things changed:

> John Davenport came in with various ideas, and started off as so many bosses do in any new company, by starting to query everyone's fees and expenses. We'd had a considerable battle over the years to get things running fairly smoothly with the accountants. . . .
>
> The differences got worse when he started telling us what to do with the cars . . . there were all sorts of tensions in the team, the mechanics were unhappy, Bill Price was unhappy, I was unhappy. . . .
>
> We even got into a situation where there was a disagreement at a service point in the Ypres rally. Now you just don't *have* disagreements on rallies—I don't think Johnstone had ever had one with the previous management in nearly ten years.
>
> I forget what the circumstances were, but the result was that we were on the very next stage, half way through the stage, when a corner came up and there was nothing coming over the intercom from the co-driver. He'd just clammed up, thinking about the row—and we went off the road. . . .

John Davenport recalls that, once appointed, he always seemed to be under pressure from his management, and from Marketing in particular. It was not easy to develop cars, and work out logical programmes, when there were so many influences from above. In the next two years:

> We continued with the Group 1 Dolomite Sprints for some time. I don't think there was any *great* pressure, but the Marketing people were keen that we should continue to do things with the Dolomites.
>
> I think we carried on with the Group 1 Dolomite for too long in rallying. What it meant was that we ended up in 1978 with two cars, with completely different mechanical parts—different gearboxes, different axles, and different engine tunes—we actually bogged ourselves down quite a lot because of this.
>
> However, I don't recall being forced to do any events that I *really* didn't want to do. We had some fairly strong National Sales companies on the continent, and we tried to spend the central budget in such a way that we were actually appearing on the territory of each of these people. My brief was: 'Look for the rally which will give us the best possible result, and also is of sufficiently high status so that we'll gain high publicity out of it.'

John also remembers that there was very little trained engineering expertise at Abingdon when he arrived:

> I hadn't really inherited a rally team, but I'd inherited a Special Tuning Department which was a commercial front, but which also went rallying . . . There was the fossilized remains of a famous 1960s rallying team, when nobody had engineers anyway. Bill Price, together with Den Green, and with Bill Burrows (who was ex-Special Tuning), they were sorting out the cars.
>
> David Wood eventually joined us, in August 1977. He arrived at about the time we started work on the TR7 V8s, but he'd actually been doing some contract work for us on the vee-8 engine before that. He was effectively our chief engineer—his title was Engineering Liaison Manager.
>
> Alan Edis and Spen King wanted me to be able to use the facilities and general knowledge which always exist in a very big company. They wanted me to form an engineering department which was able to do its own designs and make its own decisions, and to be able to liaise with other engineering departments. But there was no pressure on me to run any particular cars, because when I arrived the programmes had already been decided, and we were locked in to running 16-valve TR7s and the Dolomite Sprints for the year.

The rallying programme, in fact, was the most ambitious Abingdon had tackled for a long time, for the TR7s were scheduled to tackle no fewer than 12 events, six of them in the UK but six of them overseas, in Europe.

Although the driving strength for 1977—Brian Culcheth and Tony Pond to drive TR7s, Pat Ryan to drive Dolomites—was unchanged, there were changes made to the way the cars were built. The Broadspeed engine connection was finally severed, which allowed Abingdon to use Don Moore for all its four-cylinder engine supplies, this time with larger Weber carburettors and more power. When the time came to work on the vee-8, Moore also supplied some early examples. The rear disc brake kit was finally ready (the 1976 cars had used rear drums), and more handling improvements.

February was a month of firsts—the first-ever European rally appearance by a 'works' TR7, a victory first time out—in fact the first-ever international rally victory. It was an extremely encouraging start to the season, especially as conditions—rain, ice, mud—were about as awful as possible.

This, incidentally, was the eighth event to have been tackled by KDU 498N (one of the original 'works' TR7s)—so the old car was promptly pensioned off, and sold. Brian Culcheth's sister car KDU 497N, on the other hand, had a lot more to do before it was retired.

Tony Pond then embarked on an early-season programme where he so often nearly won an event, and had to settle for second or third place. In the very snowy Mintex rally (with its stages mostly in the fast forests of North Yorkshire) he led the event for a time, until punctures pushed him back to third place. On the island of Elba (off the coast of Italy), a mostly-tarmac/all-pace-notes event, the TR7 finished third overall. Pond then missed out on the Welsh rally, when he rolled his car, but came back with a rebuilt car to finish second on the Scottish, once again being cheated of victory by several punctures.

Brian Culcheth, on the other hand, was not having as good a season, with his cars suffering from mechanical problems. He, too, was affected by the deteriorating personal atmosphere in the team. Pat Ryan's Dolomite Sprints, however, kept going and were usually in the running for victory in the Group 1 category.

Perhaps it was unwise of the team even to enter the Mille Pistes rally of France, which was always run in very hot weather, and which used very rough and rocky stages in the hills

In spite of snowy conditions, the TR7s performed well on the Mintex rally in February 1977, taking third (Tony Pond) and 17th (Brian Culcheth, in this car).

Triumph's second successful foray into Europe, in 1977, was in Elba, where Tony Pond took third place overall on this tarmac-and-pace-notes event. (Gordon Birtwistle)

behind the Mediterranean coast close to Marseilles. Damage to the cars, rather than uncompetitiveness, was likely to be the biggest problem (shades of the Liege–Sofia–Liege rallies of the early 1960s!) In the end both cars *were* damaged, one with a broken axle, the other with a broken shock absorber which punctured the fuel tank. Fortunately for Culcheth, who had not been having a good year, this happened close to the end of a stage flying finish, and Bill Price's mechanics were able to repair the car. The result was that Culcheth was rewarded with fourth place overall.

An expedition to the Hunsruck rally of West Germany was a disappointment, but this was followed by two test sessions—one at Chobham, the other at Cadwell Park. These were more important than usual—in one respect because this was the first time a TR7 V8 rally car had been driven in anger, the other because a new tarmac handling package was being settled.

It was difficult to get a feel for the TR7 V8 at first, as John Davenport recalls:

Once we had got the 16-valve TR7 sorted out, we were able to concentrate on the TR7 V8, but it wasn't quite as simple as just taking a vee-8 engine and sticking it in. We had to tune the vee-8 engine, and nobody, to be quite honest, knew the first thing about tuning the Rover V8 within the company, and very people outside it. The last guy who had used a tuned vee-8 engine was Bill Shaw in those racing Rovers.

Once we'd got a tuned vee-8, and took it down to Wales to do a test session, back to back with a TR7 16-valve, well, no-one was very keen on it, because it wasn't a very nice car—it was an old test car. But once we started pressing the stop watch buttons, though, we began to realize that the vee-8 was the way to go.

It was the aftermath of the Cadwell Park test and the Manx rally, however, which brought all manner of simmering discontents into the open, as Brian Culcheth told me:

We did a considerable amount of testing at Cadwell Park, a circuit which might have been designed with the Isle of Man in mind. Tony and I found that the car worked best of all with a rear anti-roll bar fitted.

My Manx car was brand new, but when I took it up the road before scrutineering I thought it was lovely but it wasn't turning in quite as well as in testing. I said maybe we should change the rear roll bar—then the mechanics said there wasn't one fitted. I then asked for it to be replaced, they put it back on, and when I tried the car out it was once again just as good as when we had been testing.

That night John Davenport called me over, said he'd heard I'd asked for the roll bar to be fitted, and said that it was *he* who now made those decisions, and that it could either be taken off again, or I could go to the airport and catch the next plane home. He said the roll bar didn't work. I muttered something like: 'Well, you're the boss', but at that moment I knew I'd be leaving the team at the end of the season.

Autosport's Peter Newton, a great personal friend of Culcheth, and at one time his would-be 'ghost' in a rallying book which was never finished, wrote that:

> Leyland's problems seemed to be the talking point of the entire 'world' throughout Thursday and Friday. Team managers elsewhere were volunteering their various versions of the story, but the rumours all suggested one thing—that the cars had arrived in rather different specification from that which had been agreed upon in testing.

It was not the way for the team to prepare itself for a high-speed event, so the fact that Brian went on to achieve second place was a credit to his professionalism. Analysis of the stage times showed that the TR7 set eight fastest stage times, and 12 second fastest times. When you consider that the fastest Escorts and Vauxhalls probably had 25 bhp more power, and were at least 200 kg lighter, this was a truly remarkable achievement. At the end of the three-day event, the TR7 was only beaten by Pentti Arikkala's Vauxhall Chevette HS, the winning margin being 133 seconds.

By this stage the engines were rated at 225 bhp at 8,000 rpm, which was almost on a par with those fitted to racing Dolomites, and they were running with 5.3:1 rear differential units to allow every one of those revs to be used.

Clearly the cars were extremely effective on tarmac. Peter Newton again:

> Obviously the cars were set up very much with racing in mind. Perhaps too much so, for the effects of heavy landings were all too obvious, the cars bottoming out badly and clumsily, giving their drivers some anxious moments. The cars were without sufficient compliance and suspension travel for the rougher sections . . . on the smoother roads, particularly downhill, they were absolute magic to watch—handling like pure racers—neat, tidy and demonstrably fast.

Brian Culcheth's new TR7 was sensationally fast in the Manx rally of September 1977, this being the first special-purpose 'tarmac racer' to have come from Abingdon. Even so, disagreement over the car's specification led to Culcheth leaving the team at the end of the season. (Gordon Birtwistle)

While the TR7s made all the headlines, and caused all the controversy, Pat Ryan usually drove fast and successfully in Dolomite Sprints. In the Manx rally of 1977, Ryan and co-driver Mike Nicholson won the Group 1 category. (Gordon Birtwistle)

It appears, however, that the punishment sustained by the crashing-out over yumps may well have accounted for Tony's engine failure. The insistent battering pushed the sump guard up into the sump itself, and this action may have strangled an oil pipe which resulted in instant mechanical anarchy. Brian's car survived the rally only with strut changes, an axle change, a fractured fuel pipe, etc.—both cars suffered considerably. To survive, and stay in a rally like Corsica, they will need considerably improved reliability. . . .

Behind all this excitement with the sports cars, however, Pat Ryan drove very rapidly, and very safely, in the Dolomite Sprint Group 1 car, finishing seventh overall and beating the 'works' Vauxhall and Ford Group 1 machinery.

Newton's report had mentioned the Tour de Corse because this would be the team's next event, and its most important European outing of the year. For all those reasons, then, Brian Culcheth was astonished by his treatment for the event:

I just knew I wasn't going to get a new contract, as simple as that. I was very low in confidence. But for Corsica I thought I would have had my Manx car [Tony Pond got *his* ex-Manx car. . . .], which was a super tarmac car. But they sent me out to Corsica with an absolute wreck, the very first TR7 rally car, KDU 497N. By then the chassis was twisted so much that you didn't know if you would go left or right.

Davenport's opinion was that the Lombard-RAC rally was much tougher and more important than Corsica. Well, it was his decision. . . .

Perhaps it was inevitable that the entry in the Tour de Corse was not a success. Tony Pond's car seized its gearbox on the way to the first special stage, while Culcheth's car began to suffer from a badly slipping clutch after the first few hours:

There was an hour's servicing time at a rest halt where the mechanics wanted to change the clutch, but Davenport wouldn't let them. It was very sad that I should be treated in this way, after having such a long relationship with the company, and the mechanics.

Even so, Brian kept struggling on, with a slipping clutch, and finished 11th overall.

The last event of the rally team's season was the Lombard-RAC rally, for which no fewer

For Triumph, a two-car entry in the Tour de Corse in 1977 was almost an unmitigated disaster. Tony Pond's car retired with gearbox failure, while Brian Culcheth's car struggled to the end, with a slipping clutch. (Gordon Birtwistle)

than four cars were entered. One Abingdon-built car, for Tony Pond, was brand-new, while Safety Devices prepared a works-backed car for ex-Skoda driver Markku Saaristo to drive. The other cars were for Brian Culcheth and for Pat Ryan to drive, Pat actually getting his first (in fact his only) drive in a TR7. This would make service and support much simpler than before, which was very good news for Bill Price and the Abingdon mechanics.

By this time, relations between the team and the motoring press had reached a new low. Not only was there an embargo on informal driver interviews, or even comments, but the Public Relations approach had changed completely. In 1975 and 1976 a great deal of chest-

Pat Ryan drove a 'works' TR7 on only one occasion—using an 'ex-Pond' car on the Lombard-RAC rally of 1977. (Gordon Birtwistle)

Tony Pond/Fred Gallagher managed eighth overall in the Lombard-RAC rally of 1977, the car being set up high to give the best possible ground clearance over the rough forestry stages. (Gordon Birtwistle)

puffing and jingoism had reduced team spokesmen to something of a laughing stock—now the reaction was to say so little that everything actually began to appear sinister!

Once again, Peter Newton appeared to get it absolutely right by quoting:

> Leyland's all-TR7 team seems to have been arrived at with a secrecy which almost suggests that they did not want us to know . . . it's probably not particularly constructive to speculate on Leyland internal policies—as a spokesman threateningly remarked to us recently: '*You* could come unstuck.'

Ludicrous, wasn't it?

The TR7s struggled to be competitive on this World Championship qualifying round, for by comparison with the Escorts, Toyotas and Fiat 131 Abarths they were now underpowered and overweight. In fact a study of special stage times showed that Tony Pond only achieved one second fastest, one third fastest, two fourth fastest and two fifth fastest times, and none of the other drivers figured at all.

Nevertheless, Pond fought gamely on, and took eighth place, nearly 24 minutes behind the leaders, while Brian Culcheth ended his association with the Triumph marque when he had to retire after suffering from wheel stud breakages which left him powerless to move. Reports suggested that this was due to wheel nuts not being sufficiently tightened at service points which, to my seasoned eyes, did not look to be as smoothly and efficiently run as those of the rivals.

Markku Saaristo, who had not completed any worthwhile pre-event testing, was never on the pace, and rolled his car in the Hafren special stage, though he eventually brought a badly damaged car to the finish, while Pat Ryan's only TR7 drive was continually troubled with engine problems, and then ended when his car's gearbox failed on the M6 motorway.

As Bill Price later wrote in his book:

> That was the end of the first year of rule by 'JD'. One could only hope that the next year would be an improvement.

On the racing front, Triumph was more successful, and more consistent. Once again Broadspeed ran Dolomite Sprints in the British Saloon Car Championship. Andy Rouse was completely tied down by the troublesome Jaguar XJ5.3C race car programme in the European series, so Tony Dron was recalled to drive one car in Britain.

As Tony later wrote about this season:

We were going for outright wins. I gave up all other work to spend the days before races getting fit and concentrating on the car, practice, the start and, above all, how to beat that battalion of Capris in Class A.

Ralph was great, that way: he was always all for total commitment and real racing against the Class A cars, and never mind all that cissy stuff about driving carefully for points in Class B.

The policy was completely successful for:

With the car at the peak of its development we had a fantastic year, beating all the Class A cars to win seven of the 12 races outright: only a tyre failure in the final race, when we were leading the class by over a minute, prevented me from taking the outright title . . . Unknown. . . . Unknown to me, Ralph opted for a very soft tyre compound which just fell apart in the race.

The 1977 Broadspeed Dolomites were faster than ever before, thanks to Abingdon's imaginative homologation of yet more parts for the cars—it had been possible to get ventilated disc brakes and a close ratio gearbox added to the lists, while the British Championship regulations allowed dual-choke Weber carburettors (48DCOEs) to be fitted in place of the standard SU instruments. It was no wonder that peak power was up to about 205 bhp.

Although British Leyland, through Unipart, once again tried hard to put Dolomite Sprint F3-engined cars on to the rostrum in the British championship, it was an uphill battle, for even with Holbay preparing the engines they were not quite as powerful as the Toyota which dominated the scene. Not even Tiff Needell and Ian Taylor, in Marches, could do the trick, although Taylor recorded one victory, at Silverstone, earlier in the year.

1978

In BL's master plan, the key to Triumph's rallying activities in 1978 was the TR7 V8—but John Davenport's team was not confident that it would be available to them until the very day that homologation was granted—on 1 April 1978.

The first 'works' TR7 V8s (TR8s, really) used the 3½ litre vee-8 in this form, with twin dual-choke Weber carburettors. On these cars the inlet manifold, really meant to mate with the original GM version of the engine, was supplied by Offenhauser.

The new car's credibility was not helped, in fact, because the TR7 V8 was a model which never existed except on the FISA homologations. This story, however, started with a car to be called the TR8, which was due to go into series production at the Speke (Liverpool) plant in the winter of 1977/1978.

By 1977 the combination of the Rover vee-8 engine with the structure of the TR7, to produce the TR8, was a very open secret, especially among the motoring Press, and its launch in North America was due early in 1978. Unhappily for BL, however, just as TR8 pilot production began in November 1977 the militant workforce at Speke walked out on an indefinite strike, which would not be called off until April 1978.

To gain Group 4 homologation, BL had to convince the authorities that 500 cars had been built, but this was simply not possible for the new car at the time. However, as John Davenport reminded me:

> In those days there was no rigorous FIA inspection system. Provided that one provided a production sheet signed by an important manager, then nobody worried. . . .

A lot of fast and persuasive talking then went on, to show that the makings of well over 500 cars were either built, partly built, or stuck in the morass of the Speke strike—the result being that homologation approval *was* gained. However, by April 1978 the TR8 had not been launched, so as a compromise to keep the BL marketing people happy, it was called TR7V8 instead. In fact the TR8 production car did not go on sale in North America until mid-1979, by which time it had been in 'works' rallying use for more than a year.

Compared with 1977, however, an entirely different rallying programme was planned, for there were only two European events on the lists, with Tony Pond as the only contracted front-line driver. With both Brian Culcheth and Pat Ryan having left the team, John Davenport's plan was to hire specialists for occasional 'one-off' drives. There was no question that Tony was already one of the three best British rally drivers—Roger Clark and Russell Brooks were the other two—and he seemed to relish grappling with the TR7V8.

Before the TR7V8 could be brought into use, Pond tackled two events—the Mintex and the Circuit of Ireland—in 16-valve four-cylinder TR7s, one of which would soon be con-

By 1978 the TR7/TR7 V8 had been modified significantly, with a revised instrument layout, and now with no need of an overdrive switch on the gear-lever knob.

A famous victory for Triumph—Tony Pond and Fred Gallagher won the 24 Hours of Ypres rally outright in 1978 in this TR7 V8. SJW 540S, using fully lowered 'tarmac' suspension, was on its very first event. Note the smart new livery (blue on red).

verted to TR7V8 status. The team might as well have stayed at home, for the Mintex was a very snowy event, which led to Pond putting his car off the road on several occasions, while on the Circuit of Ireland (when the famous red/blue livery was seen for the first time) the engine blew up after a radiator hose had come adrift.

The TR7V8, complete with its close-ratio five-speed gearbox, its Rover-Triumph (as opposed to modified-Salisbury) rear axle, and all the latest development bits and pieces, made its public debut in the TV Rallysprint at Esgair Dafydd, near Llandrindod Wells. This short test showed, without question, that the TR7V8 was already a much faster car than the four-cylinder TR7 had ever been, the result being that Pond's car finished second overall, just a couple of seconds behind Hannu Mikkola's 'works' Escort RS.

The procurement problem was still serious, but easing, though close-ratio gearbox supply remained a problem. John Davenport reminded me that:

> One of the first things Bill Price alerted me to, when I joined the department, was that he had already been trying through Engineering Procurement to get close-ratio gears manufactured, for at least 18 months. It wasn't until the racing Rover programme was looming, with the same basic gearbox, that we managed to get something done, and then only by kicking up a hell of a fuss.
>
> Procurement was not favourably inclined to Motorsport, they were overloaded with work, and eventually I persuaded the company, round about 1979, to let us do our own Procurement.

This situation, I promise you, is factually reported. One wonders why a company was actually *in* motorsport if it was not prepared to provide a proper backing?

On its first rally as a homologated machine—the Granite City, based on Aberdeen—the TR7V8 won outright, but in the next two home internationals Pond retired, once with an alternator problem, the other because he rolled the car so badly that it had to be reshelled.

SJW 540S, having won in Ypres in June 1978, then went on to win in the Manx rally of September 1978. Same car, same crew—same result.

Then came the 24 hr of Ypres rally, in June, in Belgium, where Tony Pond won outright in a brand-new car. The car had a dry-sump engine, was low and stiff, rather like the 1977 Manx cars had been; looking and sounding like a racing car.

This, like the Boucles de Spa in 1977, was one of the great occasions for the team, as Ypres always drew a large and very competitive entry. Because it was an all-tarmac/all-pace notes event, Pond was able to practise carefully, and beat everyone else by three minutes.

In spite of this, however, morale at Abingdon was still low, as ex-team members have made clear, and it never seemed to make sense when changes were made from proven components, often with disastrous results. One testing change, too—the trial of Goodyear instead of Dunlop tyres—was not conclusive.

Then came the second big win of the year when Tony Pond, using his ex-Ypres 'tarmac racer', won the Manx International rally, although the Irish driver, Derek Boyd, had to retire when his engine was destroyed after a dry sump oil pump drive belt broke. By this time the cars' characteristics were well known—they could be supreme on tarmac, they had powerful and torquey engines which produced up to 280 bhp, but they continued to suffer from problems with the Rover-derived five-speed gearbox.

The team's biggest functional disaster, through no fault of its own, came in the Tour de Corse in November. Tony Pond was entered in his successful Ypres and Manx-winning V8, had practised very diligently, and was joined on this occasion by the French driver Jean-Luc Therier.

The cars could have been extremely competitive, on an event which would otherwise be dominated by the 'works' Fiat Abarth 131s. The Triumphs started with high hopes, but they were absolutely shattered when both cars were forced to retire at a very early stage with seized gearboxes. In both cases it was found that the gearbox drain plugs had been loosened and the lubricant had disappeared. As *Autosport* reported in its rallying column:

Sabotage on the TR7s. There now seems to be little doubt that both BL-entered V8-engined TR7s in Corsica were sabotaged the day before the rally. Both cars . . . were out of the event by the end of the second special stage, Therier's car with gearbox failure and Pond's car with apparent clutch failure. On the first special stage, both cars had been leaking oil from the gearbox and . . . both gearbox drain plugs were found to be only finger tight.

Many years later, in 1991, John Davenport was still not ready to use the word 'sabotage' but:

> I can't think why anyone should want to sabotage our cars. Certainly *someone* had loosened the gearbox plugs, and *someone* had made sure the oil was let out. It isn't as if we were a great threat to anybody. . . .

Which only left the Lombard-RAC rally for the team to regain its composure, and its reputation. Not only did the TR7V8s carry some commercial sponsorship—from British Airways, in a free-seats for exposure deal, no doubt, as such airline tie-ups were normally arranged—but there were three of them, Tony Pond being joined by John Haugland of Norway (who, up to then, had only been noted for driving small, slow, rear-engined Skodas) and veteran driver Simo Lampinen, with whom John Davenport had won the RAC rally as long ago as 1968 (in a Saab V4).

Tony Pond was expected to be fast and competitive, even though the TR7V8 was by no means fully sorted for forestry events; Pond had actually told pressmen that his car was 'virtually undriveable on the loose'. Right from the start, however, his rally was really ruined in Trentham Gardens when the handbrake cable linkage came adrift and jammed up a rear wheel for several minutes.

In the end it was a miracle that Tony fought his way back to fourth place overall (behind three 'works'-specification Escorts), setting eight fastest stage times, 12 second fastests, 13 third fastests and ten fourth fastest times. Perhaps he could not have expected to beat Mikkola's and Waldegard's Fords (which shared 51 fastest stage times between them), but he richly deserved to take third place. As it transpired, he was nearly 16 minutes off the pace. John Haugland brought his TR7V8 into 12th place, but Lampinen's car retired with a clutch failure.

Even before the end of what had been a difficult and very controversial year, Tony Pond

Having won two tarmac events, SJW 540S was then driven by Tony Pond in the Lombard-RAC rally of 1978, carrying British Airways sponsorship. Pond's fourth place was the best ever achieved by a TR7 or TR7 V8 in Britain's premier event.

had decided to leave the team to take up a very lucrative offer from Chrysler to drive the new Chrysler Sunbeam-Lotus. (Of this car, team manager Des O'Dell once said: 'They told me that to beat the Escorts I would have to design a better Escort. So I did . . . .'). He had been frustrated by the car's apparent lack of progress, and by the deteriorating personal atmosphere at Abingdon.

Even John Davenport admits that Pond had become frustrated by the car's lack of competitive pace in forestry events, where its traction and handling still left a lot to be desired, but he also says that:

> It wasn't such a bad career move as far as Tony was concerned. He was also offered a great deal more money to join Chrysler. However, I was delighted when, after only one year, he called me and suggested that we got together again. Was there a personality problem at Chrysler? You'd have to ask Tony about that. . . .

Although Dron, Broadspeed and the Dolomite Sprint stayed together for one final racing season in 1978, Dron has always said that the car felt worse while actually going faster—his actual comment was that engine development had taken a second off the lap times, but chassis changes (including the use of rose-jointed rear suspension) had lost half of that improvement

The 1978 engines produced about 215 bhp, but Ford's Capris had made even greater advances, the result being that the Sprint won only one race outright—the first of the season, at Silverstone—but won Class B yet again.

By this time it was clear that the Sprint had reached its limits. Since BL, and John Davenport, were concentrating completely on the TR7V8, and its rallying future, the Dolomite Sprint was therefore retired from 'works' racing, and left to the private owners for future years.

In F3, Unipart made one final attempt to turn the Dolomite Sprint into a race-winning F3 engine. There was a change of engine supply and of preparing team. Dave Price racing

The Triumph Dolomite Sprint was used in a series of F3 cars, sponsored by Unipart, in an attempt to win the British F3 series. Ian Taylor is at the wheel of this Dolomite-engined March; Tiff Needell is crouching alongside.

took over the running of the March cars, while Holbay lost the engine build and development contract to Swindon Racing Engines, whose ingenuity even stretched as far as providing different cam profiles on the same cam lobe to suit the inlet and exhaust valves respectively. That's clever, and complex, if you think about it!

Tiff Needell stayed loyal to the team throughout the season, though his original partner, Ian Taylor, was replaced by the young New Zealander, Brett Riley. Needell actually finished fourth in the British BP Super Visco F3 series—behind Nelson Piquet, Derek Warwick and Chico Serra, and sixth in the parallel Vandervell Championship. There were no victories, though Needell led races from time to time, and the cars took second and third places on occasion. Even so, Unipart, Dave Price Racing and Swindon would try again in 1979.

1979

By this time the motorsport department at Abingdon was not only looking troubled on the surface, but there was also a great deal of paddling and commotion underneath. John Davenport remembers that after the first two years in his job he no longer reported to Alan Edis but:

> He then went off to the Harvard Business School, and I reported to David Caswell (who worked for Spen King on quality matters), then there was a big change. Functionally Jaguar-Rover-Triumph was split off from Austin-Morris, but MG was still tacked on to Austin-Morris, and we were on the ground at the MG factory.

This was irritating, for it meant that theoretically Triumph's motorsport effort was now to be completely isolated from that of the Triumph marque, which meant that matters of procurement and technical co-operation were bound to become more difficult.

> Peter Frierson, who was the MG Plant Director, and I, then started a major lobby to get us transferred jointly into J-R-T, because it made sense, because our mainstream motorsport programmes were with Rover and Triumph. Eventually we got approval for that, which meant that I reported to Peter Morrow, who was the marketing director of J-R-T.
>
> It was complicated, and for some time I always seemed to be plugged into one area, then unplugged and added to another area. Motorsport was rather like an ice-hockey puck. It really wasn't until Harold Musgrove and Mark Snowdon of Austin-Rover took an interest, which must have been in 1982/1983, that things improved.

Because British rallying was, and still is for that matter, a small 'village', almost all the top-class drivers seemed to know why Tony Pond had left the team at the end of 1978. Joining a team to drive a car like the TR7V8 was one thing, but joining a team which was led by a very strong-minded director was an added complication.

John Davenport realized this, and decided to do things his own way. A promising young driver, Graham Elsmore (who had made his name driving Ford Escorts) was retained as the regular driver, while Scandinavian 'stars' such as Simo Lampinen and Per Eklund were hired for most of, but by no means all, the other events in the programme. Derek Boyd did two events in Ireland at the start of the year, in one of which he was involved in a heavy crash which injured co-driver Fred Gallagher quite severely; fortunately Fred made a complete recovery.

Once again Abingdon was faced with preparing cars for a very mixed programme—for there would be entries in Great Britain, in Southern Ireland, in Belgium, Finland and Italy during the year. By this time, too, David Wood's modification programme was in full swing, especially in relation to the engine tune:

John Davenport told me:

> 'We couldn't improve the suspension much. I think the lack of upward wheel movement was part of the problem. We couldn't get struts of the right length into the car because of the low

bonnet line. We went to enormous lengths to try to get the suspension to work better so that you could jump nicely.

'The answer, too, was that it had quite a wide track and a short wheelbase compared with the Escort, which didn't help in some conditions. It got to be quite good, but its forte was tarmac, no question.

'An awful lot was made, at the time, on the subject of driver visibility. That did have an effect, although when you look at Carlos Sainz and the modern Celica you begin to wonder if that's actually so important . . .'

When the body shells were being made at Speke, Abingdon was forced to take what it was given, but after stamping and assembly was moved to Swindon:

'We did go in for a weight reduction programme, and managed to get some of the unstressed panels—doors, bonnets, things like that—in slightly thinner steel.

Unlike some other 'works' teams in the business, BL certainly didn't adopt the policy of having two or more cars all of which carried the same registration numbers. Not only that, but a given body shell was kept going as long as it seemed to be serviceable, perhaps too long for the liking of some of the drivers. As John Davenport recalls:

'The TR7V8 had a very strong body shell. Safety Devices did our shells in those days, it was a tight shell with a lot of strength in the floorpan. We tended to keep every individual car until someone came in from the workshop and said: 'Well, we *really* ought to get rid of that one!'

A lot of work was done on the engine, as Davenport confirmed:

There was a lot to come from the engine. On a few occasions we ran with Pierburg fuel injection. Pierburg approached us, and we gave them a try. We had had rather discouraging experience of Lucas systems on the racing Jaguars; it was a relatively crude system. Lucas also had a system for the proposed USA version of the TR8 and the SDI, and that wasn't a very happy arrangement either.

'So, when Pierburg approached us and said that they'd got an electro-mechanical system, we tried it out. We got better response with that engine, but there wasn't enough development done on it. In round figures we were getting around 300 bhp.

In later years Tom Walkinshaw, at TWR, got about 360 bhp from the engines as used in Rover 3500 (SD1) race cars. I think if we'd carried on with the TR7V8, and been able to use that engine in our car, it would have made a great deal of difference to the performance.

Graham Elsmore drove 'works' Triumphs in 1979, but did not have a very successful season. Here he has the ex-Pond Ypres/Manx winning TR7 V8 well sideways on the Welsh rally, just before he went off the road and crashed the car.

This page *Once again, in the Isle of Man in 1979, Graham Elsmore showed that the TR7 V8 could be formidably effective on tarmac stages. Elsmore finished third overall. (Fred Gallagher)*

Davenport's gamble in retaining Graham Elsmore to drive in most events was not a success, and I think even the unhappy Elsmore (who never got another 'works' drive after his season with BL) would agree with this. He started 12 events but retired on seven of them; he had one victory (in a national-status event), a third place on the Isle of Man, but rarely looked as fast, or as committed, as Tony Pond had done in 1978.

In the first half of the year the cars seemed to retire, or crash, more often than they finished, but at least Per 'Pekka' Eklund was second overall in the Mintex (where the weather conditions were appallingly snowy), and won the TV Rallysprint at Esgair Dafydd.

The Pierburg fuel-injected engine made its debut on the 1000 Lakes rally, where Per

Only Per Eklund, a Swedish driver, could possibly have enjoyed conditions like this! In the 1979 Mintex, where conditions were wintry throughout, he took a magnificent second place overall. (John Davenport)

Eklund drove it, but this car suffered from power losses as the engine bay became hot, while Lampinen's carburated car suffered a broken distributor rotor arm, and eventually ran out of time after service work went on too long at one stage. It was a credit to Eklund's determination that he managed to finish eighth overall.

After the drama and controversy which surrounded the Tour de Corse, it was surprising that the TR7V8s were entered for the San Remo. After all, if certain people thought it important enough to sabotage the cars on Corsica, how much more critical would it be to ensure that they did not perform in nearby Italy? Not that it mattered in the end, because both cars retired, one with engine failure.

The season ended, as usual, with a mass entry—this time of four cars—on the Lombard-RAC rally. By this time the media was not expecting the TR7V8s to go well in loose surface conditions, especially as the cars were seen to be no more reliable than they had been at the start of the season. Although there was a modicum of British Airways support for the cars, and three cars struggled to the finish they were not impressive.

As a long-time Triumph enthusiast I naturally hoped that the cars would excel in Britain's premier rally, but I was so disappointed with their form that in my own *Autocar* report on the event, I described the TR7V8 as 'the BRM of rallying' and suggested that:

> 13th, 16th, and 17th, with the most powerful cars in the event and full works backing, is a miserable performance . . . the basic problem still seems to be the lack of top-quality drivers, and loose surface handling. Our *Autocar* testers know that the cars are already the fastest in rallying, so a search for more power (with fuel injection) seems pointless. There are promises, yet again, that everything will come right in 1980, but that, after all, will be the TR7's fifth year in rallying. . . .

The comment on top-class drivers stemmed from the fact that the American John Buffum,

In 1979 and 1980 British Leyland dabbled with a Pierbug fuel-injection installation for its TR7 V8s. It was neat, and gave more flexible power than earlier vee-8s, but it was neither totally reliable nor popular with some team members.

who had usually been thought of as a star only in North American rallying terms, drove a works-prepared but privately-entered fuel-injected TR7V8 on the event, and was faster than the 'works' drivers before he put the car off the road.

Once again Unipart made a great effort to become race winners in British F3 racing, with March 783 cars and Dolomite Sprint engines and, on balance, were more successful then ever before. Dave Price racing ran the cars, Swindon Racing Engines tuned the 16-valve Dolomites, and the cars were driven by Nigel Mansell and Brett Riley. Even with all this

San Remo rally 1979, with Per Eklund pressing hard in the dust.

Simo Lampinen of Finland, who had first driven for Triumph in 1965, also drove 'works' TR7 V8s in 1978 and 1979. (Fred Gallagher)

commitment, however, the team was not quite on a par with other teams using Toyota engines, and the cars could only finish fifth and eighth in the 20-event series.

In the Vandervell British F3 Championship, therefore, Mansell and Riley each won one race, while Mansell had two second places, and Riley finished second once. Riley's outstanding performance was victory in the European Championship race in May—when the slight lack of power was offset by his driving ability in the wet. All in all, these two were on the leader board in one of the top six positions on 23 occasions.

1980

BL Motorsport faced the 1980 season in rather a sombre mood. Money was always tight, there was no way that the cars could make a full-scale assault, or be successful, at World Championship level, and for that reason it was difficult for John Davenport to attract a 'Superstar' driver to his team. The contrast with Ford, who had not only won five World Championship rallies in 1979, but who had used Hannu Mikkola, Bjorn Waldegard, Ari Vatanen and Roger Clark as its drivers, was complete.

Davenport, however, was a resourceful and persuasive operator. For 1980, not only did he secure a sizeable budget, so that his cars could tackle a full programme, but he also persuaded Tony Pond to return to the team after a one-year 'sabbatical' with Chrysler UK, while a surprise recruit to the TR7V8 ranks was Roger Clark.

The major technical advance to the car was that a four-dual-choke Weber carburettor installation (inspired by Bill Price while David Wood was otherwise engaged with fuel injection) was developed for the vee-8 engine, this bringing the power output up to at least 300 bhp, with more torque and better fuel efficiency than before. There were still reliability problems to be solved, notably in the lubrication system, these taking ages to diagnose and cure.

Tony Pond had not had a happy year with Chrysler, where he had had many differences of opinion with team boss Des O'Dell. His return to Abingdon, therefore, where he was to be backed up by Per Eklund, was welcome. There was no question that, since 1976, Pond had been the only driver who appeared to have no fear of the TR7's sometimes self-willed character, and who had given the car its greatest successes.

Roger Clark's arrival, to tackle a British programme, was influenced by two factors. One was that Ford had just announced its withdrawal from rallying, to get on with the development of a new car. The other was that the Clark family's garage group had recently expanded to take on a BL dealership in Leicester, so the use of a TR7V8 was very wise in marketing terms.

Davenport thought long and hard before settling on a programme which included forays to Southern Ireland, Portugal, France, Belgium and Finland—yet some flights of fancy were denied him:

> Believe it or not, I actually put together a budget to do the Safari, because I felt the TR7V8 was at least as good as the Datsun 240Z had been, and where we used rough going the car seemed to soak it up very well despite the rather small amount of wheel movement.
>
> However, having put together that budget, I don't think I ever had the courage to put it forward as a serious suggestion! The TR7V8 on the Safari, well, it might not have done so badly.

Roger Clark, in fact, had rather a miserable year, for his engine blew on the first three events which he tackled, and of seven events which he started, he only finished once—in ninth place on the Scottish rally.

It was a season which threatened to be a complete disaster, for only two cars finished any events before the end of May. Entries in the important Mintex, Portugal, Circuit of Ireland, and the French Criterium Alpine events were a complete wash-out.

Although Per Eklund had finished second in the *Daily Mirror* rallysprint (Tony Pond went off the track), this had been little consolation. Pond's luck then changed when, driving an old car which Terry Kaby had had on loan during 1979, he used Michelin tyres to win the Manx stages rally. Although this was only a national-status rally, a win was a win,

Roger Clark (left) and Jim Porter alongside their Sparkrite-sponsored TR7 V8, which they used in the 1980 rally season. Roger's family business had just taken on a British Leyland franchise, so the marketing connection made a lot of sense.

Unquestionably the fastest, the best handling, and the most successful of the 'works' TR7 V8s was TUD 683T. It won two major events in 1980—Ypres and the Manx. This was the Tony Pond/Fred Gallagher combination in the Isle of Man.

and the lugubrious Pond's moustache finally began to bristle. Most importantly, the car looked, sounded, and went very well indeed.

In the rest of the year Pond then went on to win the 24 hr of Ypres, the Manx International rally and (in company with the F1 driver Alan Jones) the pre-RAC Rally rallysprint. Most significantly of all, Per Eklund finished third in the Finnish 1000 Lakes rally.

The Ypres outing, in particular, was an outstanding success for, as in 1978, the TR7V8 showed how good it really was on tarmac, and on an event where the crew had practised assiduously. It was an event in which very fast Porsches were the favourites, but Pond was in such aggressive form, and the Michelin tyres were so effective, that he ended up victorious by more than eight minutes.

Because the 1000 Lakes is such a specialized event, Davenport did not ask Tony Pond to drive. His team leader, naturally enough, was Per Eklund, but there was universal surprise when he invited veteran Timo Makinen (who had been out of top-line rallying since 1976) to drive the other car. Of all things, Timo's car ran out of fuel on a special stage, and was eliminated, but Eklund finished third overall—which was the TR7V8's best-ever World Championship rally result. Eklund's car regularly set times in the 'top six'—33 times in 47 special stages—and was beaten only by Markku Alen's Fiat 131 Abarth and Ari Vatanen's Ford Escort RS.

Only a few days later, Tony Pond boosted Abingdon's morale even further by taking TUD 683T to the Isle of Man where, in full tarmac trim, and with Michelin tyres, his TR7V8 won the Manx International rally—this being Tony's second consecutive Manx International win in Triumphs (for in 1979 he had been driving a Sunbeam-Lotus). Nor was this a walkover, for he was chased all the way home by Ari Vatanen's Rothmans Escort RS, and by Jimmy McRae's Vauxhall Chevette HSR; at four minutes, the winning margin was comfortable enough.

The last event of the year was the Lombard-RAC rally, for which no fewer than four 'works' TR7V8's were prepared—three for the regular drivers, the fourth for John Buffum of North America. Buffum, incidentally, had recently had great success in the SCCA series in the USA, where on special stage events rather like those we knew so well in the British Isles, his TR7V8 had won many events. It was a complex team operation, for Buffum's car used BF Goodrich tyres and a Pierburg injected engine, whereas the other cars used Michelins, and four-Weber engines.

It was typical 'RAC Rally luck' for Tony Pond, for on the very first special stage, through the wildlife park at Longleat House, he slid his car off the road, and rammed a wooden structure in the lions' area. Fortunately for him all the lions had been locked away for the day, but unfortunately for him one wooden bar was at such a height that his car almost slid under it, and he had to tackle the rest of the event with a crumpled roof, and with a replacement windscreen taped into position.

That, however, was only the start of Tony's problems, for he later slid off into a ditch on a stage, staying there for more than 20 minutes. Effectively, therefore, he was then last, and had to claw his way back up the lists.

All in all, this was a typically 'if only . . .' performance by the TR7V8s, for they appeared to handle a lot better than in 1979 and were driven with a great deal more determination. Per Eklund held third place for many hours until his engine blew in Kielder, John Buffum's car retired with a broken axle on the very last special stage, and Pond eventually came back to take seventh place overall. By fantasizing, and taking away the 'lost 20 minutes' from his stage penalties, I reckon that he could so easily have been in third place.

Well before the end of the year, however, BL's (and Triumph's) whole future in motor-sport was threatened by the news that top management had decided to close down the MG factory at Abingdon, and to sell it off to the highest bidder. This, at a stroke, meant that the Motorsport operation was likely to be thrown out into the street, which obliged Davenport to look around for new premises:

> I actually had to start phoning every plant manager in the BL empire, asking if, by any chance, they had about 40,000 sq.ft. of space to spare?

This, a series of budget cuts, and the requirement for John Davenport to cut back on his head-count, made everyone at Abingdon uneasy, and it soon became clear that the Triumph rally programme might have come to the end of the road:

John Davenport told me:

> I got instructions from Ray Horrocks, no less. He said that the TR7/TR8 production programme was in doubt, and that I was to stop using the cars in motorsport. The other reason, of course, was that we'd started to use the Rover 3500 in motor racing in 1979. This was already winning, and there was a lot of backing behind it.

The axe fell just days after the Lombard-RAC rally, when Davenport was instructed to 'lose' no fewer than 17 of the 47 jobs under his control—and, of course, he also had to find a new home for the surviving members. BL announcements to the media mentioned financial problems—there were many of those at the time throughout the corporation—and empha-sized that the 1981 motorsport budgets would be spent on Rover 3500 racing, and on Metro developments.

Aftermath

It was the end of the road for the TR7V8, but a lot of the experience gained could be applied to the Rover 3500. Although everyone—Davenport, Wood, Price and Pond cer-

A year after the last official 'works' Triumph had appeared, John Buffum drove a Leycare-backed ex-works car in the Lombard-RAC rally of 1981.

tainly—all think that the TR7V8 had reached the peak of its development by 1980—within, that is, the constraints of the regulations. If times had been different it might have been interesting to re-homologate the car into the Group B category for one or two seasons.

Not only would the engine have been good for a reliable 360 bhp by then, but Davenport would also have liked to homologate an alternative Getrag gearbox, to replace the Rover 77 mm box. Consider the potential of such a car on tarmac, in the Isle of Man or in pace note events in Europe?

Triumph in Motorsport—The End

When the last TR7V8 rally car was sold off, and when BL Motorsport moved to its new home at Cowley, this spelt the end of Triumph's 52-year 'works' involvement in motorsport. There was no awkward tailing off in activity, thank goodness, but a clean break. In November 1980 the team entered its last rally—and by the spring of 1981 no trace of the TR7V8s remained at Abingdon. When the department moved to Cowley in June 1981, even the memories were extinguished.

Triumph, in any case, was on the way out. The last real Triumph saloons—Dolomites and Dolomite Sprints—were built in November 1980, while the last TR7 and TR8 sports cars were produced (at the Rover factory, at Solihull) in October 1981. For the next three years the 'Triumph' badge appeared on an anglicized Honda, but this was always a sham, and seen to be so. It was a sad end to a famous name.

Appendix 1
The BRM Connection

Well before Standard-Triumph's competitions department had been set up, Sir John Black was briefly connected with motorsport—in fact with the ill-fated BRM Grand Prix car project.

The connection, as always, was a result of friendships built up over the years. Racing driver Raymond Mays, who was the 'father' of the British Racing Motors (BRM) project, had known Sir John for many years, so his approach for support paid off almost immediately.

The story really started in 1921, when the young Raymond Mays had started to campaign a Hillman 1½-litre Speed Model car in sprints and hill-climbs. At that time Capt. John Black (as he then was) was already a director of Hillman in Coventry.

Mays took his Hillman to the factory to ask for advice, met Capt. Black for the first time, and persuaded him to provide special parts for the engine. A few weeks later the modified car was raced at Brooklands for the very first time, winning one race and finishing second in another, this establishing Mays's reputation with Capt. Black!

In the 1930s, when Mays was already one of Britain's most successful racing drivers, he decided to start building his own road cars. By that time Capt. Black had become Standard's managing director, and he personally approved Mays's plan to use a Flying Standard V8 rolling chassis as the basis for his new car, though very few were ever built.

Even before the Second World War Raymond Mays had conceived an ambitious plan. In the 1930s he had seen the way that Nazi government backing of the Mercedes-Benz and Auto-Union teams had helped them dominate Grand Prix racing.

He thought the same idea could be repeated in Britain. As he later wrote in his book *BRM* (which has been out of print for many years):

> My idea was to build a team of three 1½-litre supercharged Grand Prix cars, not, as in the past, backed by a single manufacturer or private enthusiast but on an entirely new basis of co-operative endeavour and with Government support. . . . Peter [Berthon] and I formed a company called Automobile Developments Ltd. and drew up a tentative scheme for enlisting the financial support we would need.

Because of the pressures of war, no progress was made on this scheme until 1945, by which time his mechanic Ken Richardson had returned to the fold. In March 1945 Mays launched an appeal to Britain's motor industry, explaining his master plan, explaining that his company could deal with all the design work, but that he needed technical advice, priority of supply and—most important—help in terms of finance and manufacturing capability. He was looking for at least £120,000 (at 1945 prices), either as cash or in work and materials.

Right from the start, two top tycoons—Oliver Lucas of Joseph Lucas Ltd., and Alfred Owen, of the Rubery Owen organization—gave backing, though several other major manufacturers refused to help. Very quickly, it seemed, the new GP project was up and running. Then:

Standard-Triumph's chief body engineer, Walter Belgrove, was consulted about the styling of the vee-16 BRM project in 1948. This was a model of the shape which he proposed. Except for the front grille, it was accepted by BRM without modification . . . (BMIHT/Rover Group).

. . . as the support from the steel men, the accessory and part makers came rolling in, I turned to Sir John Black, head of the Standard Motor Company.

. . . we had first met a quarter of a century before, when he was Mr J. P. Black, a director of Hillmans, and I was the undergraduate owner of a Hillman sports car which I was anxious to race at Brooklands. Then, when I went to Coventry to ask Hillman's help in tuning up my car, John Black insisted in trying it out himself. . . .

This time I was again asking for help in racing—on the grand scale.

We were, in fact, old friends now. At different times we had been going to build cars together but plans had never come off. I had been to him on many projects and I do not think he ever turned me down.

He had a superb suite of rooms at the Standard offices, and to interview Sir John was like interviewing royalty. He always dressed very smartly and he had everything done in a magnificent way. When tea was brought in on a trolley the cups were really beautiful with 'J.B.' on them in gold. He was a great showman.

I outlined the proposition to Sir John and asked him if he would attend a luncheon of my supporters at the Angel Hotel, Bourne. Sir John promised to think the idea over and to consult his directors.

On the day before the lunch he telephoned me to say that he and the Standard Company were going to back us. I thanked him and asked if he would be coming to the lunch. 'No, ' answered Sir John, 'I am afraid I am too busy. But Alick Dick, my deputy Managing Director, will be there. And he is going to hand you a cheque for £5,000'.

By 1948 the British Motor Racing Research Trust had been set up, Raymond Mays was running a new Triumph 1800 Roadster, and the grandiose racing plan was well under way, with Ken Richardson acting as chief mechanic and test driver, and with a total work force of more than 30 people. Standard had not only provided the £5,000 cheque, but had already agreed to provide its toolroom facilities for the machining of engine castings—the large and complex upper and lower crankcases, and the cylinder heads.

Not only that, but the company had also provided technicians to design and install the engine test beds at Bourne with new equipment.

. . . as this testing shot of 1950 makes clear. Ken Richardson, still employed by BRM at that stage, is at the wheel. In future years, however, many more slots and scoops were needed to cool the engine.

This work was completed by the end of 1948, but completion of the first engine, then the first car, dragged on and on. At a later stage Standard was also consulted about the style of the new car—the result being that Walter Belgrove, who also styled the original Ferguson P99 four-wheel-drive racing car in later years, became involved in the development of the shape of the first vee-16 engined car. The first car was not finished until December 1949, it did not put in an appearance on a race track until mid-1950 (when it failed, on the starting line), and it was not race-reliable until 1953, by which time the GP formula for which it had been designed had been abandoned.

By that time, too, Standard had lost interest in the project.

There was one other consequence, of course. As already related, Ken Richardson left BRM at the end of 1951, which ensured that he was available to Standard-Triumph to help develop the TR2 in 1952 and 1953, and to manage the competitions department from 1954 to 1961.

Appendix 2

RESULTS

1953-1980

Between 1953 and 1980 the Triumph 'works' teams were based at several different sites, under several different managements. It is not surprising, therefore, that when researching this book I found that virtually no important archives have survived. Even so, I *think* this Results Appendix gives a complete story of the international events contested by factory-supported cars over almost 40 years. In some cases I have included references to 'works-supported' entries, either because the cars were 'works-owned', or because there was 'works involvement' in the entry itself.

My special thanks go to several of the drivers, particularly to Brian Culcheth for double-checking this list, and to Bill Price for the work he did in cataloguing the 'Abingdon' section of this Appendix.

In the immediate post-war period Standard-Triumph did not have a motorsport programme. From time to time Coventry-registered cars appeared in major rallies (such as the RAC or Monte Carlo events), but these were always standard cars which had been loaned to dealers or well-known drivers for the occasion.

The first 'works' motorsport department was set up by Ken Richardson, at Banner Lane in 1954, first of all to build the Mille Miglia car, then the Alpine rally team cars. The 'preview' of that operation was the successful TR2 high-speed demonstration on the Jabbeke road in Belgium in 1953:

Year/ Month	Event	Car	Crew (including co- driver if known)	Reg. No. (if known)	Result

Works-supported cars built by a department managed by Ken Richardson
(Located, successively, at the Banner Lane, Coventry, factory, then at the Allesley, Coventry (Service Department), and finally at Radford, Coventry.

1953 May	Jabbeke high-speed demonstration	TR2 prototype	Ken Richardson	MVC 575	124.889 mph

1954

Month	Event	Car	Drivers	Registration	Result
March	Lyons–Charbonnières Rally	TR2	G. Grant/P. Reece	OHP 676	6th Overall

(Note: This car was a factory-loaned ex-Press car, with little back-up from the factory. At the time Gregor Grant was Editor of *Autosport*)

Month	Event	Car	Drivers	Registration	Result
April	Rally Soleil Cannes	TR2	M. Gatsonides/Feurtrier	OVC 262	2nd in Class
May	Mille Miglia	TR2	M. Gatsonides/K. Richardson	OVC 276	27th Overall
June	Le Mans 24-Hour race	TR2	E. Wadsworth/B. Dickson	OKV 777	15th Overall

(Note: Though not officially a works entry, there was works assistance in the enterprise, particularly as a result of the Mille Miglia outing)

Month	Event	Car	Drivers	Registration	Result
July	Alpine Rally	TR2	M. Gatsonides/R. Slotemaker	PDU 20	6th Overall, 2nd in Class Won *Coupe des Alpes*
		TR2	K. Richardson/K. Heathcote	OVC 276	4th in Class Ret—Hub bearing.
		TR2	J. Ray/L. Mills	PDU 21	Along with Kat/Tak in a private TR2, Triumph won all possible Team Prizes
September	Tourist Trophy	6 TR2s started, including one 'works-sponsored' car: TR2	K. Richardson/B. Dickson	OKV 777	23rd Overall, 6th in Production Class. Along with two privately-prepared TR2s, won the Team Prize

1955

Month	Event	Car	Drivers	Registration	Result
January	Monte Carlo rally	Standard 10	J. Wallwork/J. Ray	PRW 893	3rd in Class
		Standard 10	K. Richardson/K. Heathcote	PRW 534	4th in Class
		Standard 10	J. Gott/R. Brookes	PRW 532	Finished 116th
		Standard 10	Ms M. Walker/Ms B. Haig	PRW 894	Finished 150th

Year/Month	Event	Car	Crew (including co-driver if known)	Reg. No. (if known)	Result
March	RAC rally	Standard 10	J. Ray/B. Harrocks	PRW 894	1st Overall
		Standard 10	K. Richardson/K. Heathcote	PRW 534	3rd Overall
		Standard 10	B. Dickson/I. Robertson	PRW 893	Finished. Won Manufacturers' Team Prize
	Lyon–Charbonnières rally	TR2	G. Grant/P. Reece	PDU 21	4th in Class
	(As in 1954, this was a 'works-loaned' car, with no other 'works' back up)				
May	Tulip rally	Standard 10	M. Gatsonides/T. St J. Foster	NRW 953	4th Overall, 1st in Class
		TR2	K. Richardson/K. Heathcote	OVC 276	2nd in Class 17th Overall
		TR2	J. Ray/J. Waddington	PDU 21	Ret—Accident
		TR2	G. Grant/S. Asbury	PKV 693	Finished 42nd
		TR2	Ms L. Grounds/Ms.C. Osborne	PKV 697	Finished
		TR2	B. Dickson/I. Robertson	PKV 698	Finished 34th
June	Le Mans 24-Hour	TR2	M. Morris-Goodall/L. Brooke	PKV 374	19th Overall
		TR2	K. Richardson/B. Hadley	PKV 375	15th Overall
		TR2	B. Dickson/N. Sanderson	PKV 376	14th Overall
August	Liege–Rome–Liege rally	TR2	K. Richardson/K. Heathcote	OVC 276	5th Overall, 1st in Class
	Tourist Trophy race	TR2	M. Gatsonides/Bourelly	RHP 557	7th Overall
September		TR2	B. Dickson/K. Richardson	PKV 376	22nd Overall
1956 January	Monte Carlo rally	Vanguard III	P. Cooper/C. Kimber	RVC 202	Finished
		Vanguard III	Ms J. Ashfield/Ms W. Clark	RVC 526	Finished
		Vanguard III	M. Gatsonides/M. Becquart	RVC 527	8th Overall, 5th in Class

Month	Rally	Car	Drivers	Reg.	Result
		Vanguard III	K. Richardson/K. Heathcote	RVC 529	Ret—suspension
		Vanguard III	P. Bolton/A. Slater/R. Richards	RVC 542	Finished
		Vanguard III	Ms. C. Osborne/Ms L.Grounds	RVC 544	Finished
		Standard 10	T. Wisdom/C. Smith	PRW 893	Finished
March	RAC rally	Standard 10	J. Wallwork/W. Cave	PRW 532	1st in Class
		Standard 10	P. Cooper/G. Holland	OOT 909	1st in Class, and 5th overall
		Standard 10	H. Rumsey/P. Roberts	PRW 894	3rd in Class
		Standard 10	P. Hopkirk/J. Garvey	PRW 893	Finished

(Note: Tom Gold, in his own Standard 10, was second in Class to Johnny Wallwork)

Month	Rally	Car	Drivers	Reg.	Result
	Lyon—Charbonnières rally	Standard 10	G. Grant/K. Heathcote	—	9th in Class
May	Tulip rally	Standard 8	P. Hopkirk/J. Garvey	SHP 876	3rd Overall
		Standard 8	D. O'M. Taylor/L. Tracey	SHP 877	4th Overall
		Standard 8	J. Wallwork/W. Bleakley	SHP 878	2nd Overall
		Standard 8	Ms C. Osborne/Ms L. Grounds	SHP 879	Ret—engine
		Standard 8	Ms J. Ashfield/ Ms M. Handley-Page	SHP 884	Finished

Also, Manufacturers' Team Prize for Wallwork, Hopkirk and Taylor

Month	Rally	Car	Drivers	Reg.	Result
	East African Safari rally	Standard Vanguard III	M. Gatsonides/V.C. Gossington	RVC 527	3rd in Class
June	Midnight Sun rally	TR3	K. Richardson/K. Heathcote	SRW 990	8th GT Category
		TR3	P. Hopkirk/W. Cave	SRW 991	5th GT Category
		TR3	Ms A. Bousquet/Ms. J.Ashfield	SRW 992	13th GT Category
July	Alpine rally	TR3	M. Gatsonides/E. Pennybacker	SRW 410	8th Overall, 1st in Class
		TR3	K. Richardson/K.Heathcote	SHP 520	Ret—lost wheel, broke transmission
		TR3	P. Hopkirk/W. Cave	SRW 991	2nd in Class
		TR3	T. Wisdom/Ms A. Wisdom	SRW 992	5th in Class

Year/Month	Event	Car	Crew (including co-driver if known)	Reg. No. (if known)	Result
					Also won Manufacturers' Team Prize
September	Leige–Rome–Leige rally	TR3	Liedgens/Rousselle	SRW 410	5th Overall, 2nd in Class
			Pierpont/Dubois	SRW 991	Retired
1957					
February	Sestriere rally	Standard 10	T. Gold/J. Waddington	—	30th Overall
March	Sebring 12 Hour race	TR3		SRW 410	
		TR3		SRW 991	
		TR3		SRW 992	

In this event three 'ex-works' 1956 TR3s were entered and driven by Standard-Triumph's North American importers. Two of the three cars finished, winning and taking second place in their class

May	Mille Miglia	TR3	Ms N. Mitchell/Ms P. Faichney	SHP 520	Ret—accident
	Tulip rally	TR3	J. Waddington/W. Cave	TRW 735	1st in Class
		TR3	T. Gold/Mrs J. Gold	TRW 736	Ret—accident
		TR3	P. Hopkirk/J. Garvey	TRW 737	3rd in Class
June	Midnight Sun rally	Standard Eight	T. Gold/W. Cave	—	95th Overall

[Three Coventry-registered Standard Eights, badged as Vanguard Juniors, started this event, registered TRW 598, TRW 599 and TRW 612. I have not been able to link crews with cars]

September	Liege–Rome–Liege rally	TR3	M. Gatsonides/A. Jetten	TRW 735	5th Overall
		TR3	B. Consten/Pichon	TRW 736	3rd Overall
		TR3	De Changy/Leikens	TRW 737	9th Overall
		TR3	Leidgens/Dubois	SHP 520	Ret—electrics. Also won Manufacturers' Team Prize

1958

Month	Event	Car	Crew	Reg.	Result
January	Monte Carlo rally	TR3A	Ms A. Soisbault/Ms P. Ozanne	VRW 219	Ret—Out of time
		TR3A	P. Hopkirk/J. Scott	VRW 220	Ret—accident
		TR3A	J. Waddington/M. Wood	VRW 221	Ret—Out of time
		TR3A	M. Gatsonides/M.Becquart	VRW 223	6th Overall, 1st in Class
		Standard 10	J. Wallwork/J. Beaumont	TRW 607	13th Overall, 2nd in Class
March	RAC rally	Standard Pennant	T. Gold/W. Cave	VWK 282	3rd Overall
		Standard Pennant	Ms A. Soisbault/Ms P. Ozanne	VWK 283	Ret—Out of time
		Standard Pennant	P. Hopkirk/J. Scott	VWK 284	Finished
		Standard Pennant	R. Goldbourn/S. Turner	VWK 285	2nd Overall, 1st in Class
		Standard 10	C. Corbishley/P. Simister	TRW 607	6th Overall, 3rd in Class —also won Manufacturers' Team Prize
	Lyons–Charbonnières rally	TR3A	Ms A. Soisbault/Ms L. Renaud	VRW 219	6th Overall, 1st in Class
		TR3A	G. Grant/G. Phillips	VWK 610	Ret—Out of time
April	Circuit of Ireland rally	TR3A	P. Hopkirk/J. Scott	VRW 220	1st Overall
		TR3A	D. Titterington/B. McCaldin	VRW 221	2nd Overall
		TR3A	E. McMillen/J. Haslett	VRW 223	Finished —also won Team Prize
	Acropolis rally	TR3A	Ms A. Soisbault/Ms L. Renaud	VRW 219	12th Overall, 3rd in Class, won *Coupe des Dames*
May	Tulip rally	TR3A	T. Gold/A. Dyke	VRW 220	Ret—mechanical
		TR3A	R. Goldbourn/S. Turner	VRW 221	1st in Class, 10th overall
		TR3A	K. Ballisat/E. Marvin	VRW 223	47th overall

Year/Month	Event	Car	Crew (including co-driver if known)	Reg. No. (if known)	Result
		TR3A	Ms A. Soisbault/Ms P.Ozanne	VWK 610	3rd in Class
		TR3A	J. Wallwork/C. Corbishley	VVC672	Ret—broken wheels
	German rally	Standard 10	M. Gatsonides/A. Jetten	VVC 675	2nd in Class
		TR3A	Ms A. Soisbault/Ms L. Renaud	VWK 610	Ladies Award
		TR3A	K. Ballisat/P. Roberts	VRW 223	10th Overall
June	Austrian Alpine rally	Standard 10	M. Gatsonides/A. Jetten	VVC 675	Gold medal finish award
July	Alpine rally	TR3A	M. Gatsonides/A. Jetten	VVC 672	Ret—brakes/crash
		TR3A	P. Hopkirk/J. Scott	VRW 220	Ret—engine
		TR3A	D. Titterington/B. McCaldin	VRW 221	3rd in Class
					8th Overall
		TR3A	K. Ballisat/A. Bertaut	VRW 223	4th Overall, 1st in Class
		TR3A	Ms A. Soisbault/Ms R. Gordine	VWK 610	Ret—accident
		Standard 10	C. Corbishley/S. Noel	TRW 607	Ret—broken engine mounting
August	Liege–Rome–Liege rally	TR3A	R. de Laganeste/R.Blanchet	VRW 221	Ret—throttle linkage
		TR3A	M. Gatsonides/R. Gorris	VRW 223	5th Overall
		TR3A	Ms A. Soisbault/Ms R. Wagner	VVC 673	Ret—engine
		TR3A	R. Leidgens/C. Dubois	VRW 219	6th Overall
Sept.	Tour de France	TR3A	Ms A. Soisbault/Ms R. Wagner	VWK 610	2nd Coupe des Dames

In addition, there was a team of works prepared British Army Motoring Association TR3As in the event, registered VVC 288, VVC 289 and VVC 290. VVC 288, driven by Lt. Col. Crosby/Major Holmes, won its class.

1959

Month	Event	Car	Crew	Reg.	Result
January	Monte Carlo rally	TR3A	M. Gatsonides/M. Becquart	WVC 247	68th Overall
		TR3A	Ms A. Soisbault/Ms N. Ferrier	WVC 248	Ret—accident
		TR3A	J. Wallwork/W. Bleakley	WVC 249	73rd overall
		TR3A	K. Ballisat/A. Bertaut	WVC 250	119th Overall
		Standard Ten	R. Gouldbourn/S. Turner	VVC 675	102nd Overall
		Standard Ten	C. Corbishley/A. Beaumont	TRW 607	Ret—accident
February	Sestriere rally	TR3A	Ms A. Soisbault/Ms Dubois	WVC 248	3rd in Class
April	Circuit of Ireland rally	TR3A	B. McCaldin/J. Scott	WVC 247	2nd in Class
		TR3A	J. Wallwork/M. Wood	WVC 249	1st in Class
		TR3A	E. McMillen/—	WVC 250	Ret—accident
	Tulip rally	TR3A	M. Gatsonides/A. Jetten	WVC 247	3rd in Class
		TR3A	Ms A. Soisbault/Ms A. Wagner	WVC 248	Ret
		TR3A	J. Wallwork/H. Brooks	WVC 249	2nd in Class, 11th Overall
		TR3A	K. Ballisat/E. Marvin	WVC 250	2nd Overall, 1st in Class
		TR3A	C. Corbishley/G. Haggie	XHP 259	4th in Class
		Standard 10	G. Grant/B. McCaldin	WWK 512	4th in Class
May	Acropolis rally	TR3A	Ms A. Soisbault/Ms R. Wagner	WVC 248	9th Overall, 1st in Class, won Ladies Award
June	Le Mans 24-Hour	TR3S	P. Jopp/R. Stoop	XHP 938	Ret—engine
		TR3S	N. Sanderson/C. Dubois	XHP 939	Ret—radiator
		TR3S	P. Bolton/M. Rothschild	XHP 940	Ret—radiator
	Alpine rally	TR3A	R. Slotemaker/de Vries	WVC 247	2nd in Class
		TR3A	Ms A. Soisbault/Ms R. Wagner	WVC 248	Ret—Broken jack
		TR3A	R. de Lageneste/H. Greder	WVC 249	1st in Class
		TR3A	K. Ballisat/A. Bertaut	WVC 250	Ret—accident
		Herald Coupé	'Tiny' Lewis/A. Nash	TL 5	9th overall

('Tiny' Lewis's Herald was his own, but had 'works' assistance)

Year/Month	Event	Car	Crew (including co-driver if known)	Reg. No. (if known)	Result
July	Monza record run	TR3A	Cambridge University AC team of drivers	XHP 259	Took Class E records at up to 4 days and 10,000 km, at up to 102 mph
August	Liege–Rome–Liege rally	TR3A TR3A	R. Slotemaker/Bootz Ms A. Soisbault/Ms R. Wagner	WVC 250 WVC 248	Ret—Out of time 4th Overall, GT Category, 1st in Class
		TR3A TR3A	C. Dubois/de Pierpoint K. Ballisat/A. Bertaut	WVC 249 WVC 247	Ret—Accident 6th Overall, GT Category, 2nd in Class
September	Tour de France (This was not Annie's usual Coventry-registered car, but her own soft-top model. The registration number is not known)	TR3A	Ms A. Soisbault/Ms M. Cancre	—	Won Ladies' award
October	German rally	TR3A TR3A	J. Sprinzel/S. Turner Ms. A. Soisbault/Ms.R. Wagner	WVC 250 WVC 248	4th in Class 3rd in Class, 2nd in Ladies' Contest
		TR3A	K. Ballisat/'Tiny'Lewis	WVC 247	2nd in Class. Also won Manufacturers' Team prize
November	RAC rally	TR3A	Ms A. Soisbault/Ms V. Domleo	WVC 248	Won Manufacturers' Team prize along with two private TR3As, 2nd in Ladies' contest
		Herald Coupé	P. Bolton/G. Shanley	YRW 266	Finished

		Car	Drivers	Registration	Result
		Herald Coupé	K. Ballisat/P. Roberts	YRW 267	Finished / Second in Manufacturers' Team prize, along with 'Tiny' Lewis's Herald (TL5)
		Herald Coupé	C. Corbishley/G. Haggie	XHP 245	Ret—Out of Time
1960 January	Monte Carlo rally	TR3A	J-J. Thuner/J. Gretener	WVC 248	Ret
		TR3A	M. Becquart/J. Blanchet	WVC 249	Ret—accident
		Herald Coupé	K. Ballisat/S. Turner	YRW 269	Ret—accident
		Herald Coupé	'Tiny' Lewis/A. Nash	YRW 268	57th Overall
		Herald Coupé	C. Corbishley/P.Roberts	XHP 245	25th Overall
		Herald Coupé	Ms A. Soisbault/Ms A. Spiers	YWK 534	Ret—out of time
		Herald Coupé	R. Slotemaker/R. Crellin	YRW 266	Ret—holed radiator
May	Tulip rally	TR3A	R. Slotemaker/R. Crellin	WVC 247	2nd in Class
		TR3A	J. Wallwork/H. Brooks	WVC 248	3rd in Class
		TR3A	K. Ballisat/S. Turner	WVC 250	4th in Class
		TR3A	Ms A. Soisbault/Ms R. Wagner	XHP 259	2nd, Ladies' Prize
		Herald Coupé	'Tiny' Lewis/A. Nash	YRW 267	1st in Class
	(Note: David Seigle-Morris, in a privately-prepared TR3A, carrying the registration number D 20, beat the 'works' team on this event)				
June	Le Mans 24-Hour race	TRS	K. Ballisat/M. Becquart	926 HP	15th Overall
		TRS	P. Bolton/N. Sanderson	927 HP	19th Overall
		TRS	L. Leston/M. Rothschild	928 HP	18th Overall
	Alpine rally	TR3A	R. Slotemaker/R. Crellin	WVC 247	Ret—axle
		TR3A	L. Leston/S. Turner	WVC 250	Ret—accident
		TR3A	D. Seigle-Morris/V. Elford	WVC 248	1st in Class
		TR3A	Ms A. Soisbault/Ms Spiers	XHP 259	Ret—lost rear wheel
		Herald Coupé	'Tiny' Lewis/A. Nash	YRW 269	Finished

Year/Month	Event	Car	Crew (including co-driver if known)	Reg. No. (if known)	Result
1961					
Jan	Monte Carlo rally (Not an official works entry, though 'works' owned and prepared)	Herald Coupé	'Tiny' Lewis/A. Nash	YRW 266	Finished
May	Tulip rally	Herald Coupé	'Tiny' Lewis/D. Stone	YRW 266	Ret—to take advantage of handicap for G. Mabbs (Herald) who therefore won outright!
	Acropolis rally	TR3A	'Tiny' Lewis/A. Nash	WVC 251	10th Overall, 2nd in Class
June	Le Mans 24-Hour race	TRS TRS TRS	K. Ballisat/P. Bolton L. Leston/R. Slotemaker M. Becquart/ M. Rothschild	926 HP 927 HP 929 HP	9th Overall 11th Overall 15th Overall —also won the Team Prize

The Competitions department closed down after the Le Mans race of 1961, Ken Richardson left the company, and the team of mechanics was dispersed.

Year/Month	Event	Car	Crew (including co-driver if known)	Reg. No. (if known)	Result
1962 *Works-supported cars built by a department managed by Graham Robson*					
May	Tulip rally	TR4 TR4 TR4	J. Sprinzel/G. Robson M. Sutcliffe/R. Fidler J-J. Thuner/J. Gretener	3 VC 4 VC 5 VC	4th in Class 3rd in Class 2nd in Class
June	Alpine rally	TR4 TR4	J. Sprinzel/W. Cave M. Sutcliffe/R. Fidler	3 VC 4VC	Ret—accident 4th Overall, 1st in Class
		TR4	J-J. Thuner/J. Gretener	5 VC	6th Overall, 2nd in Class

Month	Event	Car	Drivers	Reg	Result
		TR4	T. Wisdom/J. Uren	6 VC	7th Overall, 3rd in Class
August	Liege–Sofia–Liege rally	TR4	J. Sprinzel/W. Cave	3 VC	Ret—radiator
		TR4	M. Sutcliffe/R. Fidler	4 VC	Ret—accident
		TR4	J-J. Thuner/J. Gretener	5 VC	9th Overall
October	Geneva rally	TR4	J-J. Thuner/J. Gretener	5 VC	14th Overall, and member of winning team
November	RAC rally	TR4	J. Sprinzel/W. Cave	3 VC	15th Overall
		TR4	M. Sutcliffe/R. Fidler	6 VC	16th Overall
		TR4	J-J. Thuner/J. Gretener	5 VC	9th Overall, 3rd in Class 2nd in Manufacturers' Team Prize
		Vitesse 1600	V. Elford/M. Butler	407 VC	Ret—gearbox
1963 January	Monte Carlo rally	Vitesse 1600	J. Sprinzel/S. Actman	6001 VC	58th Overall
		Vitesse 1600	M. Sutcliffe/R. Fidler	6002 VC	76th Overall
		Vitesse 1600	V. Elford/M. Butler	6003 VC	24th Overall, 3rd in Class
		TR4	J-J. Thuner/J. Gretener	5 VC	47th Overall, 2nd in Class
April	Tulip rally	TR4	V. Elford/D. Stone	3 VC	4th in GT Category, 2nd in Class
		TR4	R. Fidler/D. Barrow	6 VC	6th in GT Category, 3rd in Class
		TR4	J-J. Thuner/J. Gretener	5 VC	10th in GT Category, 4th in Class, GT Winners, GT Category, Team Prize

Year/ Month	Event	Car	Crew (including co-driver if known)	Reg. No. (if known)	Result
June	Alpine rally	TR4	V. Elford/D. Stone	3 VC	Ret—Accident
		TR4	M. Sutcliffe/R. Fidler	6 VC	Ret—Accident
		TR4	J-J. Thuner/J. Gretener	5 VC	Ret—clutch
August	Spa–Sofia–Liege rally	TR4	R. Clark/B. Culcheth	4 VC	Ret—gearbox
		TR4	D. Grimshaw/R. Dixon	3 VC	Ret—broken jack
		TR4	J-J. Thuner/J. Gretener	5 VC	Ret—clutch
		Vitesse 2-litre prototype	V. Elford/T. Hunter	6003 VC	Ret—engine fire
October	Geneva rally	TR4	J-J. Thuner/J. Gretener	(Own car)	6th Overall, 2nd in Class
November	RAC rally	TR4	V. Elford/D. Stone	3 VC	Ret—head gasket
		TR4	R. Fidler/D. Grimshaw	6 VC	Ret—Accident
		TR4	J-J. Thuner/J. Gretener	5 VC	3rd in Class
1964					
January	Welsh rally	Spitfire	R. Fidler/J. Hopwood	412 VC	2nd Overall
April	Shell 4000 rally	TR4	J-J. Thuner/R. Fidler	CAG 410 (ex 5 VC)	
		TR4	B. Rasmussen/P. Coombe	(ex 3 VC)	
		TR4	G. Jennings/Homsey	(ex 6 VC)	Winners of GT Category Team Prize

(Three cars were re-registered CAG 408, CAG 409 and CAG 410—USA identities—for this event)

Year/ Month	Event	Car	Crew (including co-driver if known)	Reg. No. (if known)	Result
June	Le Mans 24-Hour race	Spitfire	M. Rothschild/R. Tullius	ADU 1B	Ret—accident
		Spitfire	D. Hobbs/R. Slotemaker	ADU 2B	21st Overall, 3rd in Class
	Alpine rally	Spitfire	J-L. Marnat/J-F. Piot	ADU 3B	Ret—Accident
		Spitfire	J-J. Thuner/J. Gretener	ADU 5B	Ret—accident
		Spitfire	R. Fidler/D. Grimshaw	ADU 6B	Ret—engine
		Spitfire	T. Hunter/P. Lier	ADU 7B	3rd in Class
		Spitfire	Ms V. Pirie/Ms Y. Hilton	ADU 467B	Ret—accident

Month	Event	Model	Crew	Registration	Result
August	Spa–Sofia–Liege rally	2000	J-J. Thuner/J. Gretener	ADU 425B	Ret—rear suspension
		2000	R. Fidler/D. Grimshaw	AHP 426B	Ret—rear suspension
		2000	T. Hunter/P. Lier	AHP 427B	Ret—rear suspension
September	Tour de France	Spitfire	J-J. Thuner/J. Gretener	ADU 5B	Ret—engine
		Spitfire	W. Bradley/R. Fidler	ADU 6B	Ret—engine
		Spitfire	R. Slotemaker/T. Hunter	ADU 7B	10th Overall in GT Category, 1st in Class
		Spitfire	Ms V. Pirie/Ms S. Reeves	ADU 467B	Ret—engine
October	Paris 1000 km race	Spitfire	J-F. Piot/J-L. Marnat	ADU 5B	1st in Class
	Geneva rally	Spitfire	T. Hunter/P. Lier	ADU 7B	2nd Overall, 1st in Class
		Spitfire	J-J. Thuner/J. Gretener	ADU 6B	5th Overall, 2nd in Class Won Team Prize (with private Spitfire)
November	RAC rally	2000	P. Bolton/N. Baguley	AHP 424B	Finished
		2000	J-J. Thuner/J. Gretener	AHP 425B	Ret—rear axle
		2000	R. Fidler/D. Grimshaw	AHP 426B	6th Overall, 2nd in Class
		2000	T. Hunter/P. Lier	AHP 427B	3rd in Class
		Spitfire	Ms V. Pirie/Ms S. Reeves	ADU 467B	Ret
1965					
January	Welsh rally	2000	R. Fidler/G. Robson	AHP 424B	Ret—engine
	Monte Carlo rally	Spitfire	R. Slotemaker/A. Taylor	ADU 6B	14th Overall, 2nd in Class
		Spitfire	T. Hunter/P. Lier	ADU 7B	Ret—accident
		Spitfire	S. Lampinen/J. Ahava	AVC 654B	24th Overall, 3rd in Class
		Spitfire	Ms V. Pirie/Ms S. Reeves	ADU 467B	Ret—OTL
		2000	P. Bolton/G. Shanley	AHP 425B	Ret—engine

Works supported cars built by a department managed by Ray Henderson, with Gordon Birtwistle as his deputy

Year/Month	Event	Car	Crew (including co-driver if known)	Reg. No. (if known)	Result
March	Sebring 12 Hour race	Spitfire	P. Bolton/M. Rothschild	ADU 1B	Ret—accident
			R. Tullius/C. Gates	ADU 2B	3rd in Class
			E. Barker/D. Feuerhelm	ADU 4B	2nd in Class
April	Circuit of Ireland rally	2000	R. Fidler/A. Taylor	AHP 424B	2nd in Class
May	Tulip rally	Spitfire	S. Lampinen/J. Ahava	AVC 654B	Ret—clutch
		Spitfire	Ms V. Pirie/Ms S. Reeves	ADU 467B	Finished
		2000	J-J. Thuner/J. Gretener	EHP 78C	3rd, GT Category, 1st in Class
		2000	R. Slotemaker/F. Geest	AHP 425B	Ret—engine
June	Geneva rally	Spitfire	J-J. Thuner/J. Gretener	ADU 5B	2nd in GT Category, 5th Overall, 1st in Class
	Le Mans 24-Hour race	Spitfire	S. Lampinen/J. Ahava	AVC 654B	2nd in Class
		Spitfire	D. Hobbs/R. Slotemaker	ADU 1B	Ret—Accident
		Spitfire	W. Bradley/P. Bolton	ADU 2B	Ret—Oil cooler
		Spitfire	C. Dubois/J-F. Piot	ADU 3B	14th Overall, 2nd in Class
		Spitfire	J-J. Thuner/S. Lampinen	ADU 4B	13th Overall, 1st in Class
	Gulf London rally	2000	R. Fidler/G. Robson	AHP 424B	Ret—transmission
July	Alpine rally	Spitfire	J-J. Thuner/J. Gretener	ADU 5B	2nd Prototype category
		Spitfire	R. Fidler/G. Robson	ADU 6B	Ret—broken wheel studs
		Spitfire	R. Slotemaker/A. Taylor	ADU 7B	Ret—engine
		Spitfire	S. Lampinen/J. Ahava	AVC 654B	1st, Prototype

					category
November	RAC rally	2000	J. Sprinzel/W. Cave	AHP 426B	Ret—gearbox
		2000	S. Lampinen/J. Davenport	EHP 78C	Ret—engine
		2000	J-J. Thuner/J. Gretener	AHP 424B	2nd in Class
		2000	R. Fidler/A. Taylor	AHP 426B	5th Overall, 1st in Class
		2000	J. Sprinzel/D. Benson	AHP 427B	3rd in Class
December	Welsh rally	2000	S. Lampinen/J. Davenport	EHP 78C	2nd Overall
		2000	R. Fidler/A. Taylor	AHP 426B	3rd Overall

During 1965 Bill Bradley used a Spitfire (ERW 512C) in sports car races in Britain, and in Europe. This was on loan, and prepared by the 'works' competitions department, but was not run as an official 'works' car.

1966

January	Monte Carlo rally	2000	S. Lampinen/J. Ahava	FHP 994C	Ret—gearbox
		2000	R. Fidler/A. Taylor	FHP 993C	14th Overall
		2000	J-J. Thuner/R. Gretener	FHP 992C	Ret—rear axle

(Note: FHP 991C was also prepared, but used only as a practice car for this event)

February	Swedish rally	2000	R. Fidler/A. Taylor	AHP 426B	Ret—accident

The 'works' team was then disbanded, and almost all the cars were sold off. Development continued under Ray Henderson and Gordon Birtwistle, work which was valuable when the BL Comps. Dept. started using Triumph 2.5PIs in 1969. In the meantime, there was limited activity in the UK:

April	Circuit of Ireland	2000	R. Fidler/A. Taylor	AHP 426B	4th Overall, 1st in Class
July	Gulf London rally	2000	R. Fidler/A. Taylor	FHP 993C	Ret—suspension and engine
November	RAC rally	2000	R. Fidler/A. Taylor	FHP 993C	Ret—engine
December	Welsh rally	2000	R. Fidler/A. Taylor	FHP 993C	2nd Overall

(Roy Fidler became RAC rally champion for 1966 with these, and other performances in National-status rallies)

Year/Month	Event	Car	Crew (including co-driver if known)	Reg. No. (if known)	Result
	During 1966 Bill Bradley used Triumph Spitfires (ERW 412C), then an ex-Le Mans car, very much modified, for a series of sports car races in Britain and in Europe. These were prepared and maintained in the competitions department, but were not officially 'works' cars. Also during 1966 Bill Bradley drove a Group 5 Triumph 2000 (one of the ex-Monte Carlo rally cars, with its registration identity removed), in the British Saloon Car Championship.				
1967 April	Circuit of Ireland	2000	R. Fidler/A. Krauklis	FHP 993C	5th Overall, 1st in Class
June	Gulf London rally	2000	R. Fidler/B. Hughes	FHP 993C	6th Overall
Nov	RAC rally	2.5PI prototypes	D. Hulme/G. Robson R. Fidler/A. Taylor	GVC 689D FHP 993C	(Event cancelled 12 hours before start, due to outbreak of Foot & Mouth disease)
1968	There was no actual competition, though development of a four-wheel-drive Triumph 1300 was carried out in Coventry. This was later driven successfully by Brian Culcheth in British rallycross events in 1969.				
	Works supported cars built by the British Leyland Motorsport department, based at Abingdon (near Oxford), and managed successively by Peter Browning, Basil Wales, Richard Seth-Smith, Bill Price and John Davenport				
1969 May	Austrian Alpine rally	2.5PI	P. Hopkirk/A. Nash	UJB 643G	Ret—clutch
June	Scottish rally	2.5PI	B. Culcheth/J. Syer	UJB 643G	2nd in Class

November	RAC rally	2.5PI	A. Cowan/B. Coyle	VBL 195H	11th Overall, 1st in Class
		2.5PI	B. Culcheth/J. Syer	VBL 196H	3rd in Class
		2.5PI	P. Hopkirk/A. Nash	VBL 197H	2nd in Class

Early in 1969 Brian Culcheth won rallycross events in Britain in the special four-wheel-drive 1300 4 x 4 prototype. This car was written off in a crash at Eggborough in April.

1970

April	World Cup rally	2.5PI	P. Hopkirk/A. Nash/ N. Johnston	XJB 302H	4th Overall, 2nd in Class
		2.5PI	E. Green/J. Murray/H. Cardno	XJB 303H	Ret—engine
		2.5PI	A. Cowan/B. Coyle/U. Ossio	XJB 304H	Ret—accident
		2.5PI	B. Culcheth/J. Syer	XJB 305H	2nd Overall, 1st in Class
June	Scottish rally	2.5PI	B. Culcheth/J. Syer	WRX 902H	1st Overall

1971

May	Welsh rally	2.5PI	B. Culcheth/J. Syer	XJB 305H	14th Overall
June	Scottish rally	2.5PI	B. Culcheth/J. Syer	XJB 305H	10th Overall, 2nd in Class
Sept	Cyprus rally	2.5PI	B. Culcheth/J. Syer	XJB 305H	2nd Overall

1972

April	East African Safari rally	2.5PI	B. Culcheth/L. Drews	KNW 798	13th Overall, 1st in Class
June	Scottish	Dolomite	B. Culcheth/J. Syer	CKV 2K	19th Overall, 2nd in Class
Sept	TAP rally	Dolomite	B. Culcheth/J. Syer	CKV 2K	Ret—suspension

1973
No entries

Year/Month	Event	Car	Crew (including co-driver if known)	Reg. No. (if known)	Result
1974					
March	TAP rally	Dolomite Sprint	B. Culcheth/J. Syer	FRW 812L	Ret—steering
June	24Hrs of Ypres rally	Dolomite Sprint	B. Culcheth/J. Syer	FRW 812L	Ret—brakes
July	Tour of Britain	Dolomite Sprint	B. Culcheth/R. Hutton	RDU 983M	Ret—engine fire
	Antibes–Grasse rally	Dolomite Sprint	B. Culcheth/J. Syer	FRW 812L	Ret—radiator
	Spa 24 hours race	Dolomite Sprint	A. Rouse/T. Dron	—	5th overall
September	Tourist Trophy race	Dolomite Sprint	T. Dron	—	3rd overall
November	Lombard-RAC rally	Dolomite Sprint	B. Culcheth/J. Syer	FRW 812L	Ret—accident

Starting in this season, as the status of the British rally championship improved, Triumph 'works' cars were entered in more British national-status events.

In 1974 Broadspeed of Southam prepared and ran two Dolomite Sprints (never registered) in the British Saloon Car Championship, on behalf of British Leyland; these were habitually driven by Andy Rouse and Tony Dron. Andy Rouse won the 2-litre class, and British Leyland/Triumph won the Manufacturers' Championship.

Year/Month	Event	Car	Crew (including co-driver if known)	Reg. No. (if known)	Result
1975					
May	Welsh rally	Dolomite Sprint	B. Culcheth/J. Syer	FRW 812L	11th Overall
May	Scottish rally	Dolomite Sprint	B. Culcheth/J. Syer	SOE 8M	Ret—accident
August	Tour of Britain	Dolomite Sprint	B. Culcheth/R. Hutton	RDU 983M	2nd Overall
September	Manx Trophy rally	Dolomite Sprint	B. Culcheth/J. Syer	FRW 812L	Ret—axle
	Tourist Trophy race	Dolomite Sprint	A. Rouse	—	Ret—gearbox
		Dolomite Sprint	R. Bell/Ms J. Birrell	—	4th in 2-litre Class
November	Lombard-RAC rally	Dolomite Sprint	B. Culcheth/J. Syer	RDU 983M	1st Group 1 category, 1st in Class

In this season, Broadspeed ran two Group 1 Dolomite Sprints in the British Saloon Car Championship; the cars were driven by Andy Rouse and Roger Bell. Rouse won the Championship outright, and naturally won the 2-litre class

1976

(In this season the British Rally Championship became 'Open' to all overseas drivers, which effectively made all such events of international status)

Month	Rally	Car	Driver	Registration	Result
January	Tour of Dean rally	Dolomite Sprint	T. Pond/D. Richards	RDU 983M	1st Group 1 Category
	Snowman rally	Dolomite Sprint	B. Culcheth/J. Syer	FRW 812L	5th Overall
		Dolomite Sprint	T. Pond/D. Richards	RDU 983M	Ret—engine
February	Mintex rally	Dolomite Sprint	B. Culcheth/J. Syer	FRW 812L	7th Overall
		Dolomite Sprint	T. Pond/D. Richards	RDU 983M	1st Group 1 Category
March	Granite City rally	Dolomite Sprint	B. Culcheth/J. Syer	FRW 812L	Ret—brakes
		Dolomite Sprint	T. Pond/D. Richards	RDU 983M	4th Overall, 1st in Group 1
May	Welsh	TR7	B. Culcheth/J. Syer	KDU 497N	Ret—engine
		TR7	T. Pond/D. Richards	KDU 498N	Ret—engine
	Lucien Bianchi rally	Dolomite Sprint	B. Culcheth/J. Syer	MYX 175P	Ret—engine
June	Scottish rally	TR7	B. Culcheth/J. Syer	KDU 497N	Ret—engine
		TR7	T. Pond/D. Richards	KDU 498N	Ret—engine
		Dolomite Sprint	P. Ryan/M. Nicholson	RDU 983M	5th in Class
	Tour of Britain	Dolomite Sprint	B. Culcheth/R. Hutton	MYX 175P	2nd Overall
		Dolomite Sprint	T. Pond/D. Richards	RDU 983M	Ret—engine
August	Burmah rally	TR7	B. Culcheth/J. Syer	KDU 497N	15th Overall
		TR7	T. Pond/D. Richards	KDU 498N	Ret—accident
		Dolomite Sprint	P. Ryan/M. Nicholson	RDU 983M	1st Group 1
	Manx Trophy rally	TR7	B. Culcheth/J. Syer	KDU 497N	5th Overall
		TR7	T. Pond/D. Richards	KDU 498N	3rd Overall
		Dolomite Sprint	P. Ryan/F. Gallagher	RDU 983M	9th Overall
					— cars also won Team Prize

Year/Month	Event	Car	Crew (including co-driver if known)	Reg. No. (if known)	Result
October	Lindisfarne rally	TR7	B. Culcheth/J. Syer	KDU 497N	Ret—electrics
		TR7	T. Pond/D. Richards	KDU 498N	9th Overall
		Dolomite Sprint	P. Ryan/M. Nicholson	RDU 983M	1st in Group N
	Castrol '76	TR7	B. Culcheth/J. Syer	KDU 497N	8th Overall
		TR7	T. Pond/D. Richards	KDU 498N	3rd Overall
		Dolomite Sprint	P. Ryan/M. Nicholson	RDU 983M	2nd in Group 1
Nov	Lombard-RAC rally	TR7	B. Culcheth/J. Syer	OOE 937R	9th Overall
		TR7	T. Pond/D. Richards	OOE 938R	Ret—suspension
		Dolomite Sprint	P. Ryan/M. Nicholson	RDU 983M	Ret—engine

In 1976 Broadspeed ran one Group 1 Dolomite Sprint in the British Saloon Car Championship. Andy Rouse was the driver, and he finished second in the 2-litre class.

Year/Month	Event	Car	Crew (including co-driver if known)	Reg. No. (if known)	Result
1977					
January	Tour of Dean rally	Dolomite Sprint	P. Ryan/M. Nicholson	RDU 983M	Ret—accident
February	Boucles de Spa Mintex Rally	TR7	T. Pond/F. Gallagher	KDU 498N	1st Overall
		TR7	T. Pond/F. Gallagher	OOE 938R	3rd Overall
		TR7	B. Culcheth/J. Syer	OOE 937R	17th Overall
		Dolomite Sprint	P. Ryan/M. Nicholson	RDU 983M	35th Overall
April	Circuit of Ireland Tour of Elba	Dolomite Sprint	P. Ryan/D. Gillespie	MYX 175P	Ret—accident
		TR7	T. Pond/F. Gallagher	OOE 938R	3rd Overall
		TR7	B. Culcheth/J. Syer	KDU 497N	Ret—throttle linkage
	Granite City rally	Dolomite Sprint	P. Ryan/M. Nicholson	RDU 983M	1st Group 1, 8th Overall
May	Welsh rally	TR7	T. Pond/F. Gallagher	OOE 938R	Ret—accident
		TR7	B. Culcheth/J. Syer	OOE 937R	Ret—engine
		Dolomite Sprint	P. Ryan/M. Nicholson	RDU 983M	2nd Group 1, 9th Overall
June	Scottish rally	TR7	T. Pond/F. Gallagher	OOE 938R	2nd Overall

Month	Event	Car	Drivers	Reg.	Result
		TR7	B. Culcheth/J. Syer	KDU 497N	9th Overall
		Dolomite Sprint	P. Ryan/M. Nicholson	RDU 983M	3rd Class, 12th Overall. The above cars also won the Manufacturers' Team prize
	24Hrs of Ypres rally	TR7	T. Pond/F. Gallagher	OOM 513R	Ret—engine
		TR7	B. Culcheth/J. Syer	OOE 937R	Ret—accident
July	Jim Clark rally	Dolomite Sprint	P. Ryan/M. Nicholson	MYX 175P	Ret—engine
	Mille Pistes rally	TR7	T. Pond/F. Gallagher	OOE 938R	Ret—axle
	Hunsruck rally	TR7	B. Culcheth/J. Syer	KDU 497N	4th Overall
		TR7	T. Pond/F. Gallagher	OOM 513R	Ret—accident
		TR7	B. Culcheth/J. Syer	OOM 937R	Ret—axle
September	Manx rally	TR7	T. Pond/F. Gallagher	OOM 513R	Ret—engine
		TR7	B. Culcheth/J. Syer	OOM 512R	2nd Overall
		Dolomite Sprint	P. Ryan/M. Nicholson	MYX 175P	1st Group 1, 7th Overall
November	Tour de Corse rally	TR7	T. Pond/F. Gallagher	OOM 513R	Ret—gearbox
		TR7	B. Culcheth/J. Syer	KDU 497N	11th Overall
	Lombard-RAC rally	TR7	T. Pond/F. Gallagher	OOM 514R	8th Overall
		TR7	B. Culcheth/J. Syer	OOM 512R	Ret—broken wheel studs
		TR7	M. Saaristo/I. Grindrod	SCE 645S	37th Overall
		TR7	P. Ryan/M. Nicholson	OOM 513R	Ret—gearbox

Broadspeed's racing effort consisted of a single car for the British Saloon Car Championship, driven by Tony Dron. This car won five of the 12 races outright, and easily won the 1.6–2.3-litre class, winning ten times. The Championship was scored on a 'points-success' basis, and was won by a slower, smaller-engined car.

1978

Month	Event	Car	Drivers	Reg.	Result
February	Mintex rally	TR7	T. Pond/F. Gallagher	OOM 513R	17th Overall

Year/Month	Event	Car	Crew (including co-driver if known)	Reg. No. (if known)	Result
March	Circuit of Ireland	TR7	T. Pond/F. Gallagher	SJW 533S	Ret—engine
	[The vee-8 engined car was homologated as the 'TR7 V8' before the TR8 actually went on sale. These cars were really TR8s in all but name]				
May	Welsh rally	TR7 V8	T. Pond/F. Gallagher	SJW 533S	Ret—alternator
June	Scottish rally	TR7 V8	T. Pond/F. Gallagher	SJW 533S	Ret—accident
	24Hrs of Ypres rally	TR7 V8	T. Pond/F. Gallagher	SJW 540S	1st Overall
August	Burmah rally	TR7 V8	J. Buffum/N. Wilson	OOM 512R	8th Overall
		TR7 V8	T. Pond/F. Gallagher	SJW 533S	Ret—suspension
September	Ulster rally	TR7 V8	D. Boyd/F. Gallagher	KDU 497N	Ret—engine
	Manx Trophy rally	TR7 V8	D. Boyd/R. Kernaghan	KDU 497N	Ret—engine
		TR7 V8	T. Pond/F. Gallagher	SJW 540S	1st Overall
October	Cork rally	TR7 V8	D. Boyd/R. Cole	KDU 497N	Ret—gearbox
November	Tour de Corse rally	TR7 V8	T. Pond/F. Gallagher	SJW 540S	Ret—gearbox
			J.-L. Therier/M. Vial	SJW 548S	Ret—gearbox
	Lombard-RAC rally	TR7 V8	J. Haughland/I. Grindrod	SJW 533S	12th Overall
			T. Pond/F. Gallagher	SJW 540S	4th Overall
			S. Lampinen/R. Broad	SJW 548S	Ret—clutch

Broadspeed prepared one Dolomite Sprint to compete in the British Saloon Car Championship. The driver was Tony Dron, who won one race outright, and won the Class B Championship.

Year/Month	Event	Car	Crew (including co-driver if known)	Reg. No. (if known)	Result
1979 February	Boucles de Spa	TR7 V8	J.-L. Therier/M. Vial	SJW 548S	Ret—distributor
	Galway rally	TR7 V8	D. Boyd/F. Gallagher	SJW 533S	Ret—accident
		TR7 V8	G. Elsmore/S. Harrold	UYH 863S	Ret—accident
	Mintex rally	TR7 V8	G. Elsmore/S. Harrold	SJW 540S	Ret—engine
		TR7 V8	P. Eklund/M. Broad	SJW 548S	2nd Overall

Month	Rally	Car	Driver/Co-driver	Reg.	Result
April	Circuit of Ireland rally	TR7 V8	D. Boyd/R. Kernaghan	SJW 533S	Ret—accident
		TR7 V8	G. Elsmore/S. Harrold	SJW 546S	Ret—accident
		TR7 V8	P. Eklund/H. Sylvan	TUD 682T	Ret—engine
May	Welsh rally	TR7 V8	G. Elsmore/S. Harrold	SJW 540S	Ret—accident
		TR7 V8	S. Lampinen/I. Grindrod	SJW 548S	12th Overall
		TR7 V8	P. Eklund/H. Sylvan	TUD 682T	Ret—engine
June	Scottish rally	TR7 V8	G. Elsmore/S. Harrold	SJW 540S	Ret—engine
		TR7 V8	S. Lampinen/F. Gallagher	SJW 548S	13th Overall
		TR7 V8	P. Eklund/H. Sylvan	TUD 682T	3rd Overall
August	1000 Lakes rally	TR7 V8	P. Eklund/H. Sylvan	TUD 682T	8th Overall
		TR7 V8	S. Lampinen/J. Markkanen	TUD 683T	Ret—distributor
September	Manx Rally	TR7 V8	G. Elsmore/F. Gallagher	SJW 546S	3rd Overall
	San Remo rally	TR7 V8	S. Lampinen/F. Gallagher	TUD 683T	Ret—suspension
		TR7 V8	P. Eklund/H. Sylvan	XJO 414V	Ret—engine
November	Lombard-RAC rally	TR7 V8	G. Elsmore/S. Harrold	SJW 546S	16th Overall
		TR7 V8	T. Kaby/B. Rainbow	UYH 863S	Ret—engine
		TR7 V8	S. Lampinen/F. Gallagher	TUD 683T	17th Overall
		TR7 V8	P. Eklund/H. Sylvan	XJO 414V	13th Overall
1980					
February	Galway rally	TR7 V8	R. Clark/J. Porter	SJW 546S	Ret—engine
	Mintex rally	TR7 V8	R. Clark/J. Porter	SJW 546S	Ret—engine
March	Rally of Portugal	TR7 V8	T. Pond/F. Gallagher	HRW 250V	Ret—engine
		TR7 V8	P. Eklund/H. Sylvan	HRW 251V	Ret—fuel pump
April	Circuit of Ireland rally	TR7 V8	R. Clark/J. Porter	XJO 414V	Ret—engine
May	Criterium Alpin rally	TR7 V8	T. Pond/F. Gallagher	TUD 683T	Ret—engine
	Welsh rally	TR7 V8	P. Eklund/H. Sylvan	TUD 686T	Ret—alternator
		TR7 V8	R. Clark/J. Porter	XJO 414V	Ret—fuel pump
June	Scottish rally	TR7 V8	T. Pond/F. Gallagher	HRW 250V	4th Overall

Year/Month	Event	Car	Crew (including co-driver if known)	Reg. No. (if known)	Result
	24Hrs of Ypres rally	TR7 V8	P. Eklund/H. Sylvan	HRW 251V	Ret—engine
		TR7 V8	R. Clark/J. Porter	XJO 414V	9th Overall
		TR7 V8	T. Pond/F. Gallagher	TUD 683T	1st Overall
		TR7 V8	P. Eklund/H. Sylvan	TUD 686T	Ret—engine
August	1000 Lakes rally	TR7 V8	P. Eklund/H. Sylvan	TUD 686T	3rd Overall
		TR7 V8	T. Makinen/E. Salonen	HRW 251V	Ret—fuel
September	Manx trophy rally	TR7 V8	T. Pond/F. Gallagher	TUD 683T	1st Overall
		TR7 V8	R. Clark/J. Porter	XJO 414V	Ret—axle
Nov	Lombard-RAC rally	TR7 V8	P. Eklund/H. Sylvan	TUD 686T	Ret—engine
		TR7 V8	J. Buffum/J. Grindrod	HRW 251V	Ret—axle
		TR7 V8	R. Clark/N. Wilson	XJO 414V	Ret—engine
		TR7 V8	T. Pond/F. Gallagher	JJO 931W	7th Overall

The last event contested by 'works' Triumphs was the Lombard-RAC rally of 1980. After this the BL Motorsport Department turned to running Rover 3500s, and began to develop the MG Metro 6R4 Group B project.

The last 'Triumph' badged cars—Acclaims, which were slightly re-developed Honda Ballades—were built in 1984.

Index